Introduction to Fiber Optics

LIGHTWAVE MICROSYSTEMS
2911 Zanker Road
San Jose, CA 95134

Introduction to Fiber Optics

John Crisp

 Newnes

Newnes
An imprint of Butterworth-Heinemann
Linacre House, Jordan Hill, Oxford OX2 8DP
225 Wildwood Avenue, Woburn, MA 01801-2041
A division of Reed Educational and Professional Publishing Ltd

℞ A member of the Reed Elsevier plc group

OXFORD   AUCKLAND   BOSTON
JOHANNESBURG   MELBOURNE   NEW DELHI

First published 1996
Reprinted 1997, 1998, 1999

British Library Cataloguing in Publication Data
A catalogue record for this book is available from the British Library

Library of Congress Cataloguing in Publication Data
A catalogue record for this book is available from the Library of Congress

ISBN 0 7506 2467 1

Typeset by Co-publications
Printed and bound in Great Britain by
MPG Books Ltd, Bodmin, Cornwall

FOR EVERY TITLE THAT WE PUBLISH, BUTTERWORTH-HEINEMANN
WILL PAY FOR BTCV TO PLANT AND CARE FOR A TREE.

Contents

Preface

An increasing proportion of the world's communications are carried by fiber optic cables. It has spread quietly, almost without being noticed into every situation in which information is being transmitted whether it is within the home hi-fi system, cable television or telecommunication cables under the oceans.

The purpose of this book is to provide a worry-free introduction to the subject. It starts at the beginning and does not assume any previous knowledge of the subject and, in gentle steps, it introduces the theory and practical knowledge that is necessary to use and understand this new technology.

In learning any new subject jargon is a real problem. When the words are understood by all parties they make an efficient shorthand form of communication. Herein lies the snag. If not understood, jargon can create an almost impenetrable barrier to keep us out. In this book jargon is introduced only when required and in easily digested snacks.

John Crisp

1

Optic fiber and light — a brilliant combination

The starting point

For thousands of years we have used light to communicate. The welcoming camp fire guided us home and kept wild animals at bay. Signal bonfires were lit on hilltops to warn of invasion. Even in these high-tech days of satellite communications, ships still carry Aldis lamps for signaling at sea, signaling mirrors are standard issue in survival packs.

It was a well known 'fact' that, as light travels in straight lines, it is impossible to make it follow a curved path to shine around corners.

Boston, Mass., USA, 1870. An Irish physicist by the name of John Tyndall gave a public demonstration of an experiment which not only disproved this belief but gave birth to a revolution in communications technology.

His idea was very simple. He filled a container with water and shone a light into it. In a darkened room, he pulled out the bung. The light shone out of the hole and the water gushed out.

It was expected that the light would shine straight out of the hole and the water would curve downwards — as shown in Figure 1.1.

But what actually happened is shown in Figure 1.2.

The light stayed inside the water column and followed the curved path. He had found a way to guide light!

The basic requirements still remain the same today — a light source and a clear

Figure 1.1
What was
expected to
happen

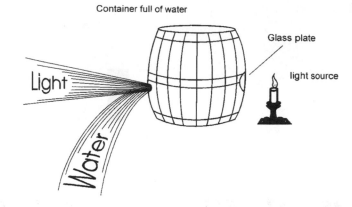

Container full of water

Glass plate

light source

Light

Water

Figure 1.2
What actually
happened

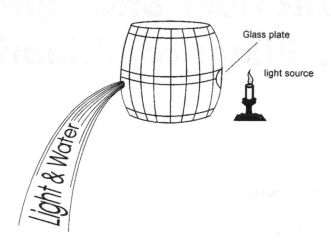

Glass plate

light source

Light & Water

material (usually plastic or glass) for the light to shine through.
The light can be guided around any complex path as in Figure 1.3.

Figure 1.3
Light can go
anywhere

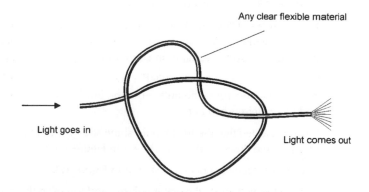

Any clear flexible material

Light goes in

Light comes out

Being able to guide light along a length of optic fiber has given rise to two
distinct areas of use, light guiding and communications.

Light guiding

There are many applications of light guiding — and more being devised every day.

Here are a few interesting examples.

Road signs

A single light source can be used to power many optic fibers. This technique is used in traffic signs to indicate speed limits, lane closures etc. The light source is built into a reflector. The front face of the reflector can then be covered with the ends of a whole series of optic fibers. These optic fibers convey the light to the display board where they can be arranged to spell out the message (Figure 1.4).

Figure 1.4

Each fiber goes to one of the 'lights'

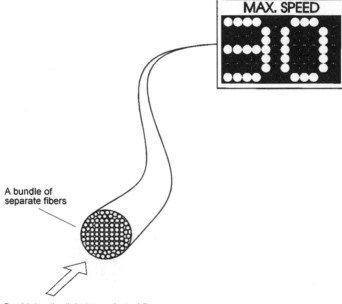

MAX. SPEED

A bundle of separate fibers

By shining the light into selected fibers we can write any message that we like

This method has the advantage that, since the whole display is powered by one electric light bulb, we never get the situation where the message can be misread — it is either all there or it's all missing.

The same method can be used to illuminate instrument panels. Instead of a series of separate light bulbs, one of which is bound to be unlit at any critical time, the whole display is powered by a single bulb — with a spare one ready to switch in when required.

Endoscopes

As the light travels down the fiber, light rays get thoroughly jumbled up. This means that a single fiber can only carry an average value of the light that enters it. To convey a picture along a single fiber is quite impossible. To produce a picture, a large number of optic fibers must be used in the same way that many separate points of light, or pixels, can make an image on a cathode ray tube.

This is the principle of an endoscope used by doctors to look inside of us with the minimum of surgery (Figure 1.5). This is a bundle of around 50,000 very thin

Figure 1.5

An endoscope

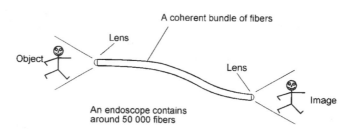

fibers of 8 μm (315 millionths of an inch) diameter, each carrying a single light level. An endoscope is about one meter in length with a diameter of about 6 mm or less. For illumination, some of the fibers are used to convey light from a 300 watt xenon bulb. A lens is used at the end of the other fibers to collect the picture information which is then often displayed on a video monitor for easy viewing.

To rebuild the image at the receiving end, it is essential that the individual fibers maintain their relative positions within the endoscope otherwise the light information will become scrambled. Bundles of fibers in which the position of each fiber is carefully controlled are called *coherent bundles* (Figure 1.6).

Figure 1.6

A coherent bundle (top) gives a good image. If it's not coherent (bottom) the image is scrambled

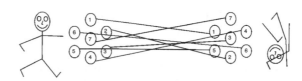

Hazardous areas

If we have a tank containing an explosive gas, a safe form of illumination is essential. One solution is to use a light source situated a safe distance away from the tank, and transmit the light along an optic fiber. The light emitted from the end of the fiber would not have sufficient power to ignite the gas.

Fire escapes

In many cases, the main danger with fire is not heat or flames but smoke. We are forced to crawl along on our hands and knees to avoid inhaling the smoke. Optic fibers are now being used to provide an illuminated strip along the lower edges of corridors to guide us safely to the exit.

Communications

In 1880, only four years after his invention of the telephone, Alexander Graham Bell used light for the transmission of speech. He called his device a *Photophone*. It was a tube with a flexible mirror at its end. He spoke down the tube and the sound vibrated the mirror. The modulated light was detected by a photocell placed at a distance of two hundred meters or so. The result was certainly not hi-fi but the speech could at least be understood.

Following the invention of the ruby laser in 1960, the direct use of light for communication was re-investigated. However the data links still suffered from the need for an unobstructed path between the sender and the receiver. Nevertheless, it was an interesting idea and in 1983 it was used to send a message, by Morse code, over a distance of 240 km (150 miles) between two mountain tops.

Enormous resources were poured into the search for a material with sufficient clarity to allow the development of an optic fiber to carry the light over long distances.

The early results were disappointing. The losses were such that the light power was halved every three meters along the route. This would reduce the power by a factor of a million over only 60 meters (200 feet). Obviously this would rule out long distance communications even when using a powerful laser. Within ten years however, we were using a silica glass with losses comparable with the best copper cables.

The glass used for optic fiber is unbelievably clear. We are used to normal 'window' glass looking clear but it is not even on the same planet when compared with the new silica glass. We could construct a pane of glass several kilometers thick and still match the clarity of a normal window. If water were this clear we would be able to see the bottom of the deepest parts of the ocean.

We occasionally use plastic for optic fiber but its losses are still impossibly high for long distance communications but for short links of a few tens of meters it is satisfactory and simple to use. It is finding increasing applications in hi-fi systems, and in automobile control circuitry.

On the other hand, a fiber optic system using a glass fiber is certainly capable of carrying light over long distances. By converting an input signal into short flashes of light, the optic fiber is able to carry complex information over distances of more than a hundred kilometers without additional amplification. This is at least five times better than the distances attainable using the best copper coaxial cables.

The system is basically very simple: a signal is used to vary, or modulate, the light output of a suitable source — usually a laser or an LED (light emitting

5

diode). The flashes of light travel along the fiber and, at the far end, are converted to an electrical signal by means of a photo-electric cell. Thus the original input signal is recovered (Figure 1.7).

Figure 1.7

A simple fiber optic

system

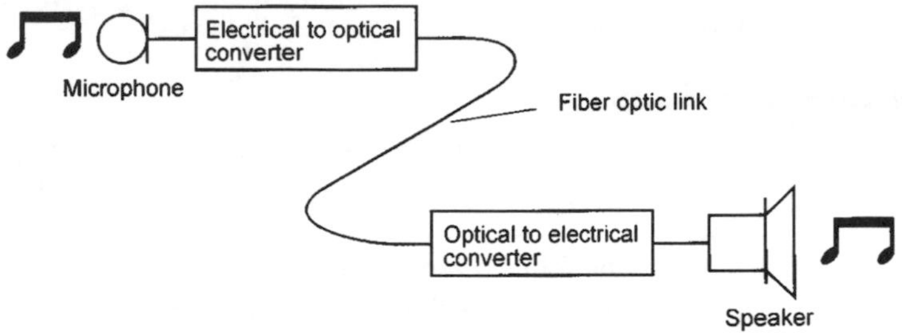

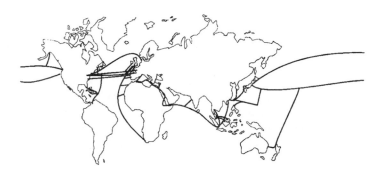

Sending information by light has proved so successful that over 90% of all long distance telephone calls are now carried on optic fibers as well as an increasing proportion of television signals. Some of the main international routes are shown in Figure 1.8.

Figure 1.8

Some of the main

fiber routes

Terminology

A brief note on some terms: optic fiber, fiber optics and fiber.

> ▷ *optic fiber* is the transparent material, along which we can transmit light.

> ▷ *fiber optics* is the system, or branch of engineering concerned with using the optic fibers. Optic fiber is therefore used in a fiber optic system.

> ▷ *fiber* is a friendly abbreviation for either, so we could say that fiber is used in a fiber system.

Quiz time 1

In each case, choose the best option.

1   **A transparent material along which we can transmit light is called:**
   (a) a fiber optic
   (b) a flashlight
   (c) an optic fiber
   (d) a xenon bulb

2   **A simple fiber optic system would consist of:**
   (a) a light source, an optic fiber and a photo-electric cell
   (b) a laser, an optic fiber and an LED
   (c) a copper coaxial cable, a laser and a photo-electric cell
   (d) an LED, a cathode ray tube and a light source.

3   **Optic fiber is normally made from:**
   (a) coherent glass and xenon
   (b) copper
   (c) water
   (d) silica glass or plastic

4   **It is *not* true that:**
   (a) endoscopes use coherent bundles of fibers
   (b) silica glass is used because of its clarity
   (c) a photocell converts light into electric current
   (d) plastic fiber is normally used for long distance communications

5   **The number of fibers in a typical endoscope is about:**
   (a) 1870
   (b) 300
   (c) 50 000
   (d) 60

2

What makes the light stay in the fiber?

Refraction

Imagine shining a flashlight. The light waves spread out along its beam. Looking down and seeing the wave crests it would appear as shown in Figure 2.1.

Figure 2.1

The wavefronts become straighter

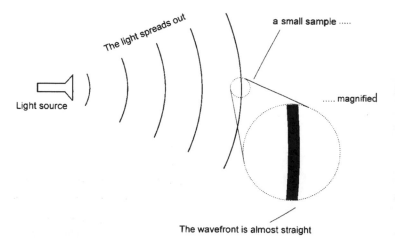

The wavefront is almost straight

As we move further from the torch, the wavefront gets straighter and straighter.

At a long distance from the light source, the wavefront would be virtually straight.

In a short interval of time each end of the wavefront would move forward a set distance.

If we look at a single ray of light moving through a clear material the distance advanced by the wavefront would be quite regular as shown in Figure 2.2.

Figure 2.2

The wavefront

moves forward

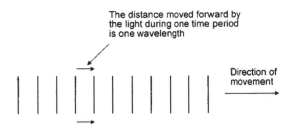

The distance moved forward by the light during one time period is one wavelength

Direction of movement

There is a widely held view that light always travels at the same speed. This 'fact' is simply not true. The speed of light depends upon the material through which it is moving. In free space light travels at its maximum possible speed, close to 300 million meters or nearly eight times round the world in a second.

When it passes through a clear material, it slows down by an amount dependent upon a property of the material called its *refractive index*. For most materials that we use in optic fibers, the refractive index is in the region of 1.5.

So:

$$\text{Speed of light in the material} = \frac{\text{speed of light in free space}}{\text{refractive index}}$$

Units

As the refractive index is simply a ratio of the speed of light in a material to the speed of light in free space, it does not have any units.

Using the example value of 1.5 for the refractive index, this gives a speed of about 200 million meters per second. With the refractive index on the bottom line of the equation, this means that the lower the refractive index, the higher would be the speed of light in the material. This is going to be vital to our explanation and is worth emphasizing:

lower refractive index = higher speed

Let's have a look at a ray of light moving from a material of high refractive index to another material with a lower index in which it would move faster. We can see that the distances between the successive wave crests, or the wavelength, will increase as soon as the light moves into the second material.

9

Now, the direction that the light approaches the boundary between the two materials is very significant. In Figure 2.3 we choose the simplest case in which the light is traveling at right angles to the boundary.

Figure 2.3

The light changes its speed

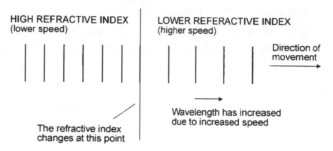

HIGH REFRACTIVE INDEX
(lower speed)

LOWER REFERACTIVE INDEX
(higher speed)

Direction of movement

The refractive index changes at this point

Wavelength has increased due to increased speed

We will now look at a ray approaching at another angle. As the ray crosses the boundary between the two materials, one side of the ray will find itself traveling in the new, high velocity material whilst the other side is still in the original material. The result of this is that the wavefront progresses further on one side than on the other. This causes the wavefront to swerve. The ray of light is now wholly in the new material and is again traveling in a straight line albeit at a different angle and speed (Figure 2.4).

Figure 2.4

The light is refracted

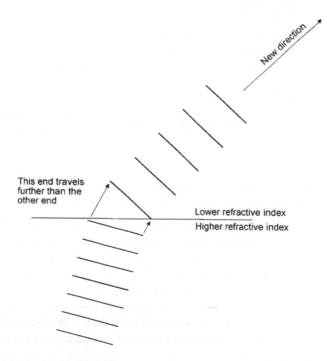

New direction

This end travels further than the other end

Lower refractive index
Higher refractive index

The amount by which the ray swerves and hence the new direction is determined by the relative refractive indices of the materials and the angle at which the ray approaches the boundary.

Snell's law

The angles of the rays are measured with respect to the *normal*. This is a line drawn at right angles to the boundary line between the two refractive indices. The angles of the incoming and outgoing rays are called the angles of *incidence* and *refraction* respectively.

These terms are illustrated in Figure 2.5.

Figure 2.5

The names of the parts

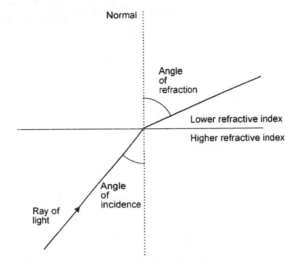

Notice how the angle increases as it crosses from the higher refractive index material to the one with the lower refractive index.

Willebrord Snell, a Dutch astronomer discovered that there was a relationship between the refractive indices of the materials and the sine of the angles. He made this discovery in the year 1621.

Snell's law states the relationship as:

$$n_1 \sin\phi_1 = n_2 \sin\phi_2$$

Where: n_1 and n_2 are the refractive indices of the two materials, and $\sin\phi_1$ and $\sin\phi_2$ are the angles of incidence and refraction respectively.

There are four terms in the formula so provided that we know three of them, we can always transpose it to find the other term.

We can therefore calculate the amount of refraction that occurs by using Snell's law.

A worked example

Calculate the angle shown as ϕ_2 in Figure 2.6.

The first material has a refractive index of 1.51 and the angle of incidence is 38° and the second material has a refractive index of 1.46.

Starting with Snell's law:

11

Figure 2.6

An example using
Snell's law

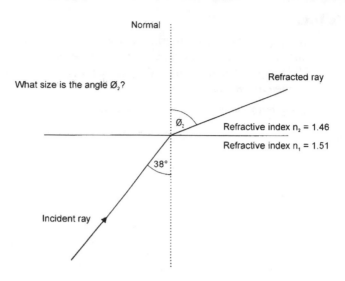

$$n_1 \sin\phi_1 = n_2 \sin\phi_2$$

We know three out of the four pieces of information so we substitute the known values:

$$1.51 \sin 38° = 1.46 \sin\phi_2$$

Transpose for $\sin\phi_2$ by dividing both sides of the equation by 1.46. This gives us:

$$\frac{1.51 \sin 38°}{1.46} = \sin\phi_2$$

Simplify the left hand side:

$$0.6367 = \sin\phi_2$$

The angle is therefore given by:

$$\phi_2 = \arcsin 0.6367$$

So:

$$\phi_2 = 39.55°$$

Critical angle

As we saw in the last section, the angle of the ray increases as it enters the material having a lower refractive index.

As the angle of incidence in the first material is increased, there will come a time when, eventually, the angle of refraction reaches 90° and the light is refracted along the boundary between the two materials. The angle of incidence which results in this effect is called the critical angle.

We can calculate the value of the critical angle by assuming the angle of refraction to be 90° and transposing Snell's law:

$$n_1 \sin\phi_1 = n_2 \sin 90°$$

As the value of $\sin 90°$ is 1, we can now transpose to find $\sin\phi_1$, and thence ϕ_1, (which is now the critical angle):

$$\phi_{critical} = \arcsin\left(\frac{n_2}{n_1}\right)$$

A worked example

A light ray is traveling in a transparent material of refractive index 1.51 and approaches a second material of refractive index 1.46. Calculate the critical angle.

Using the formula for the critical angle just derived:

$$\phi_{critical} = \arcsin\left(\frac{n_2}{n_1}\right)$$

Put in the values of the refractive indices:

$$\phi_{critical} = \arcsin\left(\frac{1.46}{1.51}\right)$$

Divide the two numbers:

$$\phi_{critical} = \arcsin(0.9669)$$

So:

$$\phi_{critical} = 75.2°$$

Total internal reflection

The critical angle is well-named as its value is indeed critical to the operation of optic fibers.

At angles of incidence less than the critical angle, the ray is refracted as we saw in the last section.

However, if the light approaches the boundary at an angle greater than the critical angle, the light is actually reflected from the boundary region back into the first material. The boundary region simply acts as a mirror. This effect is called total internal reflection (TIR).

Figure 2.7 shows these effects.

The effect holds the solution to the puzzle of trapping the light in the fiber. If the fiber has parallel sides, and is surrounded by a material with a lower refractive index, the light will be reflected along it at a constant angle – shown as ø in the example in Figure 2.8.

Any ray launched at an angle greater than the critical angle will be propagated along the optic fiber. We will be looking at this in more detail in Chapter 4.

Figure 2.7

Total internal

reflection

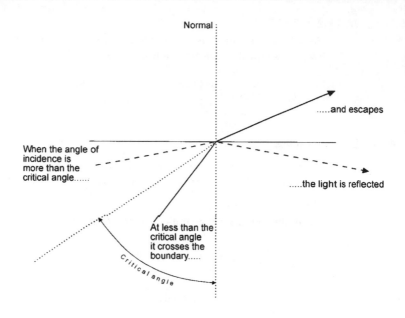

Normal

.....and escapes

When the angle of
incidence is
more than the
critical angle......

.....the light is reflected

At less than the
critical angle
it crosses the
boundary.....

Critical angle

Figure 2.8

Light can bounce its

way along the fiber

The angle stays the same all the way along

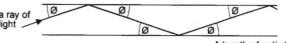

a ray of
light

A length of optic fiber

Quiz time 2

In each case, choose the best option.

1   The speed of light in a transparent material:

(a) is always the same regardless of the material chosen

(b) is never greater than the speed of light in free space

(c) increases if the light enters a material with a higher refractive index

(d) is slowed down by a factor of a million within the first 60 meters

2   A ray of light in a transparent material of refractive index 1.5 is approaching a material with a refractive index of 1.48. At the boundary, the critical angle is:

(a) 90°

(b) 9.4°

(c) 75.2°

(d) 80.6°

3   If a ray of light approaches a material with a greater refractive index:

(a) the angle of incidence will be greater than the angle of refraction

(b) TIR will always occur

(c) the speed of the light will increase immediately as it crosses the boundary

(d) the angle of refraction will be greater than the angle of incidence

4   If a light ray crosses the boundary between two materials with different refractive indices:

(a) no refraction would take place if the angle of incidence was 0°

(b) refraction will always occur

(c) the speed of the light will not change if the incident ray is traveling along the normal

(d) the speed of light never changes

5   The angle ? in Figure 2.9 has a value of:

(a) 80.6°

(b) 50°

(c) 39.3°

(d) 50.7°

Figure 2.9

Question 5:

calculate the angle ?

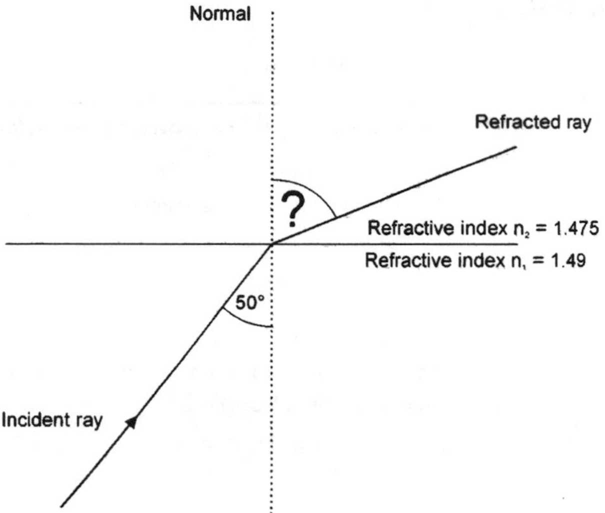

Normal

Refracted ray

?

Refractive index n_2 = 1.475

Refractive index n_1 = 1.49

50°

Incident ray

3

The choice of frequency

Electromagnetic waves

Radio waves and light are electromagnetic waves. The rate at which they alternate in polarity is called their frequency (f) and is measured in Hertz (Hz), where 1 Hz = 1 cycle per second.

The speed of the electromagnetic wave (v) in free space is approximately 3×10^{8} ms^{-1}. The term ms^{-1} means meters per second.

The distance traveled during each cycle, called the wavelength (λ) can be calculated by the relationship:

$$\text{wavelength} = \frac{\text{speed of light}}{\text{frequency}}$$

In symbols, this is:

$$\lambda = \frac{v}{f}$$

By transposing we get the alternative forms:

$$\text{frequency} = \frac{\text{speed of light}}{\text{wavelength}} \left(f = \frac{v}{\lambda} \right)$$

and:

$$\text{speed of light} = \text{frequency} \times \text{wavelength} \left(v = \lambda f \right)$$

Some useful multiples

Here are some common multiples used in fiber optics:

M Mega = 1000 000 = 1×10^6

k kilo = 1000 = 1×10^3

m milli = $\dfrac{1}{1000}$ = 1×10^{-3}

μ micro = $\dfrac{1}{1\ 000\ 000}$ = 1×10^{-6}

n nano = $\dfrac{1}{1\ 000\ 000\ 000}$ = 1×10^{-9}

p pico = $\dfrac{1}{1\ 000\ 000\ 000\ 000}$ = 1×10^{-12}

Note: micron is the previous name for the micrometer 1×10^{-6} m and is still commonly used within the fiber optics industry.

Electromagnetic spectrum

In the early days of radio transmission when the information transmitted was mostly restricted to the Morse code and speech, low frequencies (long waves) were used. The range of frequencies able to be transmitted, called the *bandwidth*, was very low. This inevitably restricted us to low speed data transmission and poor quality transmission (Figure 3.1).

Figure 3.1

Fiber optics use visible and infrared light

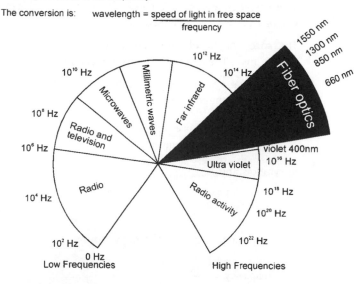

In fiber optics, we find it more convenient to use the wavelength of the light instead of the frequency.

The conversion is: wavelength = $\dfrac{\text{speed of light in free space}}{\text{frequency}}$

As time went by, we required a wider bandwidth to send more complex information and to improve the speed of transmission. To do this, we had to increase the frequency of the radio signal used. The usable bandwidth is limited by the frequency used — the higher the frequency, the greater the bandwidth.

When television was developed we again had the requirement of a wider bandwidth and we responded in the same way — by increasing the frequency. And so it went on.

More bandwidth needed? Use a higher frequency. For something like sixty years this became an established response — we had found the answer!

Until fiber optics blew it all away.

The early experiments showed that visible light transmission was possible and we explored the visible spectrum for the best light frequency to use.

The promise of fiber optics was the possibility of increased transmission rates.

The old solution pointed to the use of the highest frequency but here we met a real problem. We found that the transmission losses were increasing very quickly. In fact the losses increased by the fourth power. This means that if the light frequency doubled, the losses would increase by a factor of 2^4 or 16 times.

We quickly appreciated that it was not worth pursuing higher and higher frequencies in order to obtain higher bandwidths if it meant that we could only transmit the data over very short distances.

The bandwidth of a light based system was so high that a relatively low frequency could be tolerated in order to get lower losses and hence more transmission range. So we explored the lower frequency or red end of the visible spectrum and then even further down into the infrared.

And that is where we are at the present time.

Infrared light covers a fairly wide range of wavelengths and is generally used for all fiber optic communications. Visible light is normally used for very short range transmission using a plastic fiber.

Windows

Having decided to use infrared light for (nearly) all communications, we are still not left with an entirely free hand. We require light sources for communication systems and some wavelengths are easier and less expensive to manufacture than others. The same applies to the photodetectors at the receiving end of the system.

Some wavelengths are not desirable: 1380 nm for example. The losses at this wavelength are very high due to water within the glass. It is a real surprise to find that glass is not totally waterproof. Water in the form of hydroxyl ions is absorbed within the molecular structure and absorbs energy with a wavelength of 1380 nm. During manufacture it is therefore of great importance to keep the glass as dry as possible with water content as low as 1 part in 10^9.

It makes commercial sense to agree on standard wavelengths to ensure that equipment from different manufacturers is compatible. These standard wavelengths are called *windows* and we optimize the performance of fibers and light sources so that they perform at their best within one of these windows (Figure 3.2).

The 1300 nm and 1550 nm windows have much lower losses and are used for long distance communications. The shorter wavelength window centered

Figure 3.2

The infrared windows used in fiber optics

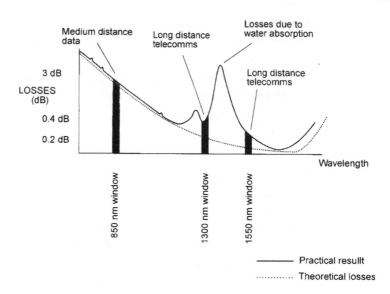

around 850 nm has higher losses and is used for shorter range data transmissions and local area networks (LANs), perhaps up to 10 km or so. The 850 nm window remains in use because the system is less expensive and easier to install.

Quiz time 3

In each case, choose the best option.

---

1   The common windows used in fiber optic communications are centered on wavelengths of:

(a)1300 nm, 1550 nm and 850 nm

(b) 850 nm, 1500 nm and 1300 nm

(c) 1350 nm,1500 nm and 850 nm

(d) 800 nm, 1300 nm and 1550 nm

---

2   A wavelength of 660 nm is often used for visible light transmission. The frequency of this light in free space would be:

(a) 660×10^{-9} Hz

(b) 4.5×10^{14} Hz

(c) 300×10^{8} Hz

(d) 45×10^{12} Hz

---

3   In free space, light travels at approximately:

(a) $186\ 000$ ms^{-1}

(b) 3×10^{9} ms^{-1}

(c) 300 ms^{-1}

(d) 0.3 meters per nanosecond

---

4   The window with the longest wavelength operates at a wavelength of approximately:

(a) 850 nm

(b) 1550 μm

(c) 1350 nm

(d) 1.55 μm

---

5   The 850 nm window remains popular because it:

(a) uses visible light and this allows plastic fibers to be used

(b) the fiber is less expensive to install and has lower losses than any other windows

(c) the system is less expensive and easier to install

(d) allows higher data transmission rates

4

Propagation of light along the fiber

Have a look at Figure 4.1. What path would be taken by the incoming ray? Would it finish up at A or B?

Figure 4.1

Does the ray go to
A or B?

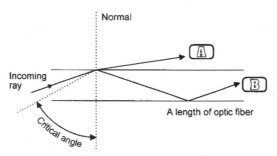

In Chapter 2 we saw that light approaching the outer edge of the optic fiber at an angle greater than the critical angle would be reflected back into the fiber and would bounce its way along as in the ray path B.

This diagram suggests that a useful transmission system can be built from a simple length of clear glass. Unfortunately, it is not that straightforward, as soon as we touch the fiber, we transfer some grease from our skin onto its surface. This causes a problem.

Problem

The contamination on the surface changes the refractive index of the material surrounding the glass. Previously, it was air which has a refractive index of 1 and it is now grease which, in common with all other materials, has a refractive index greater than 1.

This will locally increase the critical angle and some of the light will now find itself approaching the surface at an angle less than the new critical angle. It will then be able to escape (Figure 4.2).

Figure 4.2

Any contamination

causes a power

loss

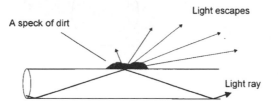

Dirt, grease, rain, in fact any contamination will allow leakage of the light. The whole situation would be impossible. We would not even be able to support the fiber, let alone keep it perfectly clean.

For every problem there is a solution.

Solution

Cover the fiber with another layer of glass as shown in Figure 4.3.

The original glass, called the core, now has a new layer, the cladding, added

Figure 4.3

How to keep the

core clean

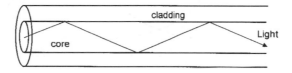

around the outside during manufacture. The core and the cladding form a single solid fiber of glass.

The optic fiber is very thin. Typical dimensions are shown in Figure 4.4.

As mentioned earlier, a μm is a millionth of a meter and was previously referred to as a micron (a term we still hear). A typical core size of 50 μm is about the thickness of one page of this book.

Our everyday experience with glass tells that it is very brittle. We cannot bend a piece of window glass — it just shatters.

With optic fiber however, this is not the case. We can easily bend it in a full circle as can be seen in Figure 4.5.

So, why the difference?

We shouldn't really ask 'why can we bend a glass fiber?' but rather 'why isn't all glass that flexible?'

When we manufacture glass, its surface is perfectly smooth until we damage it. Glass is very delicate and the normal manufacturing process is enough to cause damage. Even simply touching its surface with a finger is enough. The result is

Figure 4.4

A typical size of fiber

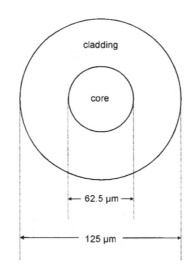

cladding

core

62.5 μm

125 μm

Figure 4.5

Bendy glass – shown full size. The bend can be as tight as that on the right before it breaks

microscopic cracking of the surface (Figure 4.6).

When we stretch it or start to bend it, we cause lines of stress to occur in the material. The stress lines are concentrated around the end of the crack (Figure 4.7). The stress builds up until it exceeds the strength of the material which responds by relieving the stress by extending the crack a little further.

This concentrates the stress even more and the material again responds by extending the crack. In this manner the crack propagates through the material a very high speeds and the glass shatters (Figure 4.8). In glass the crack propagation speed is about 1700 ms^{-1} (4000 mph). An optic fiber would break in less than a tenth of a microsecond and a window pane in a leisurely 3 microseconds.

The difference then, between the optic fiber bending in a circle and the sheet of window glass is the presence of those minute surface scratches. To protect the optic fiber from surface scratches, we add a layer of soft plastic to the outside of the cladding. This extra layer is called the primary buffer (sometimes simply called the *buffer* and is present only to provide mechanical protection and has nothing to do with light transmission (Figure 4.9).

During manufacture, the core and cladding are made simultaneously and immediately after the glass has cooled sufficiently the primary buffer is added so that it can be stored on a drum without damage.

Figure 4.6

A very small

surface flow

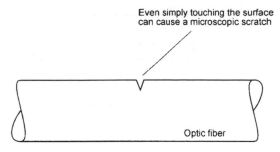

Even simply touching the surface
can cause a microscopic scratch

Optic fiber

Figure 4.7

A build-up of stress

around the tip of

the crack

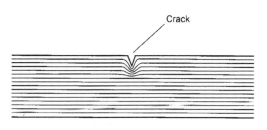

Crack

Figure 4.8

The crack

propagates

Stress is relieved by extending
the crack but this causes more
stress and makes the situation
even worse

Typical dimensions are shown in Figure 4.10.

To enable the glass to be used industrially, it has a further protective layer added. Or several layers. The choice of material for the outer layer, called the jacket, depends upon the use to which the cable is to be put. Chapter 8 looks at the choices in more detail.

Three points which are important to appreciate:

 ▷ the optic fiber is solid, there is no hole through the middle
 ▷ the buffer and the jacket are only for mechanical protection
 ▷ the light is transmitted through the core but to a small extent, it travels in the cladding and so the optical clarity of the cladding is still important.

So, why does the light enter the cladding?

If the angle of incidence is greater than the critical angle, the light ray is refracted back into the first material (TIR).

To be refracted it must enter the cladding, as shown in Figure 4.11.

We can now see that an opaque cladding would prevent the ray from being propagated along the fiber since the light would not be able to pass through the cladding.

Figure 4.9

The completed

optic fiber

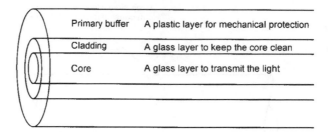

Primary buffer	A plastic layer for mechanical protection
Cladding	A glass layer to keep the core clean
Core	A glass layer to transmit the light

Figure 4.10

The primary buffer
diameter is a
standard size of
many fibers

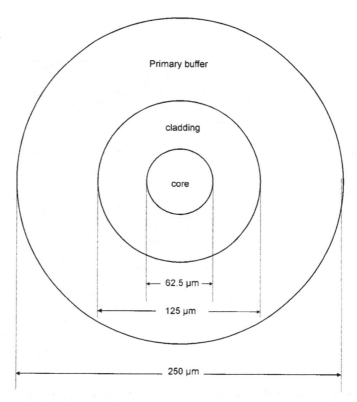

Primary buffer

cladding

core

|← 62.5 µm →|

|← 125 µm →|

|← 250 µm →|

Figure 4.11

Light enters
cladding during
reflection

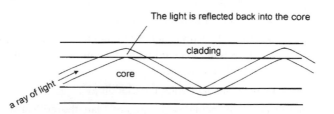

The light is reflected back into the core

cladding

core

a ray of light

Getting the light into the fiber

When we shine light along a fiber, it shines out of the far end.

We can see the light spreading out from the end of the fiber in Figure 4.12, and we can calculate the angle at which it spreads out by Snell's law.

Figure 4.12

See how the light spreads out from the end

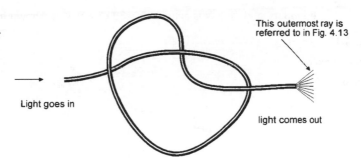

This outermost ray is referred to in Fig. 4.13

Light goes in

light comes out

Here is the situation

The ray shown has bounced its way along the fiber at the critical angle. It leaves the fiber to be the one on the outermost edge (Figure 4.13).

Figure 4.13

All light paths are reversible

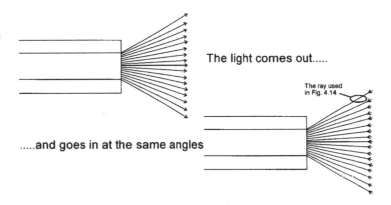

The light comes out.....

The ray used in Fig. 4.14

.....and goes in at the same angles

Let's assume the same values that we have used previously.

> ▷ core refractive index = 1.5
> ▷ cladding refractive index is 1.48

This gives an approximate critical angle (as in Figure 4.14) of:

$$\phi_{crit} = \frac{1.48}{1.5} = 0.9866 = 80.6°$$

Figure 4.14

The ray travels along the fiber at the critical angle

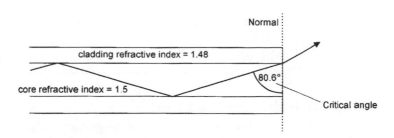

Normal

cladding refractive index = 1.48

core refractive index = 1.5

80.6°

Critical angle

As the angle is measured from the normal, the angle between the cladding boundary and the ray is actually 90–80.6° = 9.4° as shown in Figure 4.15.

Figure 4.15

The new angle of incidence

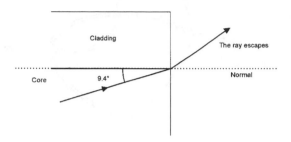

The refractive index of the air is 1 (near enough), so we can calculate the angle at which the ray leaves the fiber by applying Snell's law once again.

Remember that the normal is the line at right angles to the change in the refractive index so it is now horizontal as in Figure 4.15.

Here we go:

$$n_1 \sin\phi_1 = n_2 \sin\phi_2$$

We know values for three out of the four terms:

- ▷ n_1 = refractive index of the core = 1.5
- ▷ $\sin\phi_1$ = sine of the angle of incidence (9.4°)
- ▷ n_2 = refractive index of air = 1.0
- ▷ $\sin\phi_2$ = sine of the angle of refraction

It doesn't matter whether we use n_1 and $\sin\phi_1$ to refer to the core or to the air. The good news is that the result will come out the same whichever way we do it so this is one part we can't get wrong.

Load the numbers that we know in to the formula:

$$n_1 \sin\phi_1 = n_2 \sin\phi_2$$

We get:

$$1.5\sin9.4° = 1.0\sin\phi_2$$

So:

$$1.5 \times 0.1637 = \sin\phi_2$$

And:

$$0.245 = \sin\phi_2$$

So:

$$\phi_2 = 14.18°$$

We have now calculated the angle at which the light spreads out as 14.18° not a very large angle but typical for a glass fiber (Figure 4.16).

Incidentally, plastic fibers have a greater angle, around 27°.

Figure 4.16

The maximum

angle of

divergence

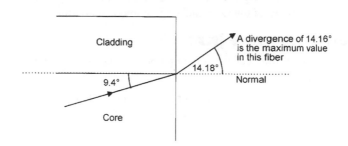

Figure 4.17
All light paths are
reversible

Since light direction is reversible, this 14.18° is also the angle at which light can approach the core and be propagated along the fiber (Figure 4.17).

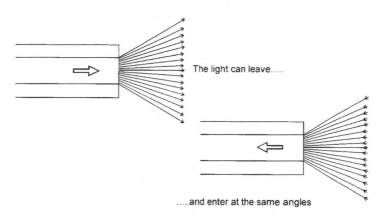

The optic fiber is circular and therefore this angle is applicable in two dimensions and would look like a cone as shown in Figure 4.18.

Figure 4.18
An acceptance
angle of 14.18°

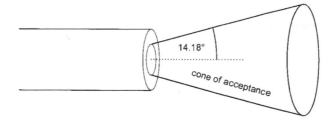

Cone of acceptance

The cone of acceptance is the angle within which the light is accepted into the core and is able to travel along the fiber.

Note

The angle is measured from the axis of the cone so the total angle of convergence is actually twice the stated value.

Calculating the cone of acceptance was no easy job and it would be nice to find a more straightforward way of finding it. Luckily there is one and it involves a property of the fiber called the numerical aperture.

Numerical aperture (NA)

The numerical aperture of a fiber is a figure which represents its light gathering capability.

We have just seen that the acceptance angle also determines how much light is able to enter the fiber and so we must expect an easy relationship between the numerical aperture and the cone of acceptance as they are both essentially measurements of the same thing.

The formula for the numerical aperture is based on the refractive indices of the core and the cladding.

There is no fun to be had in deriving it, so here it is:

$$NA = \sqrt{n^2_{core} - n^2_{cladding}} \quad \text{(no units)}$$

and here is the short cut to the acceptance angle:

$$\text{acceptance angle} = \sin^{-1} NA$$

Let's try the short cut and see how it works out using our previous figures of n_{core} = 1.5, and $n_{cladding}$ = 1.48.

This would give a numerical aperture of:

$$NA = \sqrt{1.5^2 - 1.48^2}$$
$$= 0.244$$

resulting in an acceptance angle of $\sin^{-1} 0.244$, or 14.12° (Figure 4.19).

Figure 4.19

Acceptance angle
= $\sin^{-1}$ NA

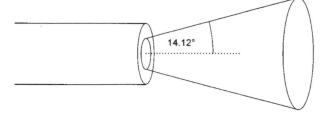

14.12°

A couple of calculator notes

▷ rounding off on the calculator has introduced a slight error in the final figure for the acceptance angle but it doesn't matter

▷ $\sin^{-1}$ on some calculators may be shown as arcsin or inv sin

If a core of a fiber is only 50 μm thick (and many are less than this) and it only accepts light within its cone of acceptance whose typical value is 12°, it would accept only a very small proportion of the light if the end on the fiber were simply held up against a light source, as shown in Figure 4.20.

Figure 4.20

A lens is often used to concentrate the light

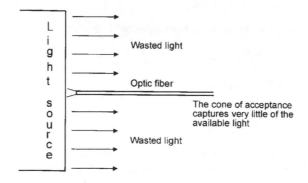

To reduce this problem, lenses are normally employed to focus the light on to the end of the fiber.

What happens to the light which approaches the fiber outside of the cone of acceptance?

Let's follow a ray and find out.

Here is the situation. The exact angle chosen is not important since all angles respond in the same way.

Choosing the previous values of refractive index allows us to know the cone angle without doing all the calculations again (Figure 4.21).

Figure 4.21

The acceptance angle is copied from Figure 4.19

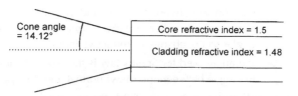

We choose an angle, anything over 14.12°, let's say 30° as in Figure 4.22, and now apply Snell's law:

$$n_1 \sin\phi_1 = n_2 \sin\phi_2$$
$$n_{air} \sin\phi_{air} = n_{core} \sin\phi_{core}$$

Figure 4.22

We choose an angle outside the cone

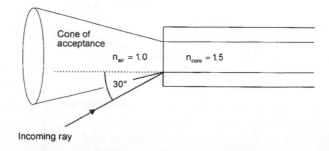

Put in all the values that we know:

$$1.0\sin30° = 1.5\sin\phi_{core}$$

Transpose, by dividing both sides by 1.5:

$$\frac{1.0\sin30°}{1.5} = \sin\phi_{core}$$

$$0.3333 = \sin\phi_{core}$$

So:

$$\phi_{core} = \sin^{-1}0.3333 = 19.47°$$

The light ray continues across the core until it finds itself at the boundary between the core and the cladding as in Figure 4.23.

Now we have to be careful. It is easy at this stage to make a mistake.

Figure 4.23

The light gets into the core

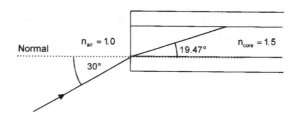

There are two problems

▷ the normal between the core and the cladding is now vertical. We must remember that the normal is always drawn at right angles to the change in refractive index

▷ the angle used for Snell's law is the angle of incidence. This is the angle between the (new) normal and the approaching ray.

In our situation shown in Figure 4.24, the ray and the two normals form a right angled triangle. In any triangle, the internal angles add up to 180°. If we subtract the 90° for the right angle and the 19.47° that we already know about, we find the new angle of incidence is:

$$180° - 90° - 19.47° = 70.53°$$

Figure 4.24

Changing the normal

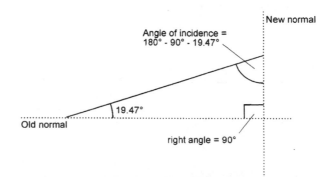

The ray will now either be reflected off the core/cladding boundary or it will penetrate the cladding.

To decide, we need to know the critical angle between the core and the cladding.

From Chapter 2 we may recall that the critical angle can be found from:

$$\phi_{crit} = \sin^{-1}\left(\frac{n_2}{n_1}\right)$$

Or, in this case:

$$\phi_{crit} = \sin^{-1}\left(\frac{n_{cladding}}{n_{core}}\right)$$

Putting in the figure gives:

$$\phi_{crit} = \sin^{-1}\left(\frac{1.48}{1.5}\right)$$

Simplifying:

$$\phi_{crit} = \sin^{-1}(0.9867)$$
$$\phi_{crit} = 80.64°$$

The ray is therefore approaching the boundary at an angle less than the critical angle and will penetrate into the cladding.

Figure 4.25

It approaches the cladding...

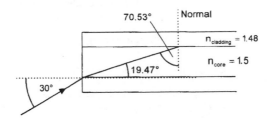

Actually, we already knew this would happen because the original ray was approaching at an angle outside of the cone of acceptance and it is only the rays inside of the cone angle that will be propagated down the core.

So now, out comes Snell's law again to sort out the situation shown in Figure 4.26:

$$n_1\sin\phi_1 = n_2\sin\phi_2$$
$$n_{core}\sin\phi_{core} = n_{cladding}\sin\phi_{cladding}$$

Put in the numbers that we know about:

$$1.5\sin70.53° = 1.48\sin\phi_{cladding}$$

Divide by 1.48:

Figure 4.26

...and into the cladding

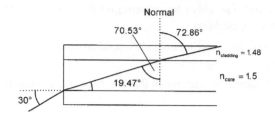

$$\frac{(1.5)(0.9428)}{1.48} = \sin\phi_{\text{cladding}}$$

Simplifying:

$$0.9556 = \sin\phi_{\text{cladding}}$$

From this, the angle cladding can be found:

$$\phi_{\text{cladding}} = \sin^{-1}0.9556$$

So:

$$\phi_{\text{cladding}} = 72.86° \text{ (give or take a bit for rounding off errors)}$$

The saga continues:

The ray now reaches the next boundary, between the cladding and the air. Will it escape into the air or will it be reflected back into the fiber?

The answer will again depend on how the angle of incidence compares with the critical angle.

Figure 4.27 shows part of the previous diagram.

Figure 4.27

Part of Figure 4.26

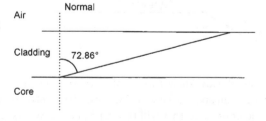

The angle between the core/cladding boundary and the previous normal is 90° and, as we know part of the angle is 72.86°, the other part must be 90°– 72.86° = 17.14° (Figure 4.28).

We now have a right angled triangle from which we can calculate the angle of incidence to be:

$$\text{angle of incidence} = 180° - 17.14° - 90° = 72.86°.$$

Notice that this is the same angle as the angle of refraction from the last boundary. This just happens to be the case whenever the two boundaries are parallel to each other.

Figure 4.28

Angle of incidence in the cladding

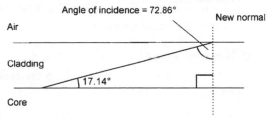

Now we check out the critical angle:

$$\phi_{critical} = \sin^{-1}\left(\frac{n_{air}}{n_{cladding}}\right)$$

$$\phi_{critical} = \sin^{-1}\left(\frac{1.0}{1.48}\right)$$

$$\phi_{critical} = \sin^{-1}0.6757$$

$$\phi_{critical} = 42.5°$$

So, where does the ray go?

The angle of incidence is greater than the critical angle and so total internal reflection (TIR) will occur. Whenever light reflects off a straight surface it leaves at the same angle as it approached at so we will have the situation shown in Figure 4.29.

Figure 4.29

...and back into the fiber

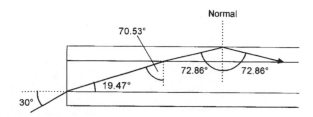

The ray is now moving back into the fiber.

Where would it go? Have a guess.

To save a lot of tedious calculations, the answer is that it will propagate down the fiber using the whole of the fiber and not just the core. All the angles are simply repeats of those we have already calculated.

Figure 4.30 shows how it would look.

Figure 4.30

The light is quickly attenuated

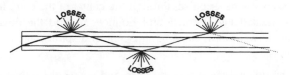

Surface contamination or the lack of clarity of the buffer, will cause serious losses along the outside of the cladding. The ray will be attenuated and will die out very quickly over the first few meters.

Names given to different rays

We have seen that rays approaching from within the cone of acceptance are successfully propagated along the fiber.

The position and the angle at which the ray strikes the core will determine the exact path taken by the ray. There are three possibilities, called the *skew*, *meridional* and the *axial* ray as shown in Figure 4.31. If light enters a fiber from a practical light source, all three rays tend to occur as well as those outside of the cone of acceptance as we have just investigated.

Figure 4.31

The skew ray does not pass through the center (top); the meridional ray passes through the center (middle); the axial ray stays in the center all the time (bottom)

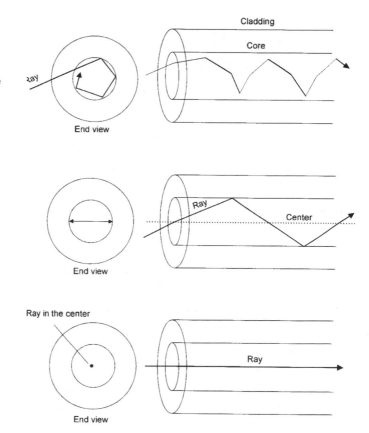

The skew ray never passes through the center of the core. Instead it reflects off the core/cladding interface and bounces around the outside of the core. It moves forward in a shape reminiscent of a spiral staircase built from straight sections.

The meridional ray enters the core and passes through its center. Thereafter, assuming the surfaces of the core are parallel, it will always be reflected to pass through the center. The meridional ray in the end view of Figure 4.31 just happens to be drawn horizontally but it could occur equally well at any other angle.

The axial ray is a particular ray that just happens to travel straight through the center of the core.

Quiz time 4

In each case, choose the best option.

1   **No material could have a refractive index of:**

(a) 1.5

(b) 1.3

(c) 1.1

(d) 0.9

2   **The ray enters the optic fiber at an angle of incidence of 15° as shown in Figure 4.32. The angle of refraction in the core would be:**

(a) 8.3°

(b) 10.14°

(c) 75°

(d) 15.54°

Figure 4.32

Question 2:

calculate the angle
of refraction

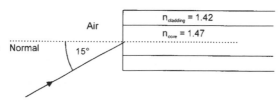

3   **The primary buffer:**

(a) reflects the light back into the core to allow propagation of light down the optic fiber

(b) must have a lower refractive index than the core

(c) is a plastic layer around the cladding to protect the optic fiber from mechanical damage

(d) must be airtight to prevent loss of the core vacuum

4   As the meridional ray is propagated along the optic fiber it:

(a) travels in a sort of spiral shape

(b) stays in the center of the fiber

(c) passes repeatedly through the center of the core

(d) is reflected off the inside surface of the primary buffer. This is called TIR

5   If the refractive index of the core of an optic fiber was 1.47 and that of the cladding was 1.44, the cone of acceptance would have an angle of approximately:

(a) 17.19°

(b) 72.82°

(c) 78.4°

(d) 34.36°

5

Decibels

To read technical information, to make sense of catalogs or to compare specifications we need an understanding of *decibels*. This chapter restricts itself to decibels as required for the field of fiber optics.

The decibel is a logarithmic unit

What is a logarithmic unit?

The word *logarithm*, even when abbreviated to the more friendly *log*, is still somewhat frightening. It instills a sense of unease. Don't worry. It's all bluff. It's actually quite simple.

Look at this:

$$10 \times 10 = 100$$

So far so good.

We can use powers of ten to write this in another form:

$$10^2 = 100$$

This is simple enough, so where is the problem? There isn't one really — except in the wording.

We can write:

- ▷ $10^2 = 100$,
- ▷   or say 10 squared is equal to 100,
- ▷   or even 10 to the power of 2 is 100.

But how would we describe the number 2 in this situation?

It is called the logarithm, or *log* of 100. It is the number to which 10 must be *raised* to equal 100.

As $10^2 = 100$ and the log of 100 is 2, and $10^3 = 1000$ and the log of 1000 is 3, it follows that the log of any number between 100 and 1000 must be between 2 and 3. We cannot work them out for ourselves so we must use a calculator.

For the log of 200

On the calculator:

> ☞   enter 200

> ☞   press the log button

and the answer 2.301029996 appears. So, ignoring some of the decimal places, we can say that 2.301 is the log of 200 or $10^{2.301} = 200$.

To perform a multiplication like $100 \times 1000 = 100{,}000$ we could add the logs of the numbers to be multiplied:

log of $100 = 2$

log of $1000 = 3$

log of $100\ 000 = 2 + 3 = 5$ or $100 \times 1000 = 10^5$

Similarly:

$$\frac{100000}{1000} = 100$$

or, by subtracting logs:

$5 - 3 = 2$

or:

$$\frac{100000}{1000} = 10^2$$

Summary

> ☞   to multiply, add logs

> ☞   to divide, subtract logs

Use of decibels in fiber optic circuits

We use decibels to compare the power coming out of a circuit or part of a circuit to the power level at the input. So basically it is an output power/input power comparison.

The decibel is a logarithmic unit and obeys the same rules as logs.

The formula is :

$$\text{power gain in decibels} = 10\log\left(\frac{\text{power}_{out}}{\text{power}_{in}}\right) \text{dB}$$

Note the abbreviation for decibels: small *d*, capital *B*, never put an *s* on the end.

An amplifier has a higher output power than its input power so it is said to have a power gain. This is the case in Figure 5.1:

$$\text{gain} = 10\log\left(\frac{\text{power}_{out}}{\text{power}_{in}}\right) \text{dB}$$

Figure 5.1

What is the gain in decibels?

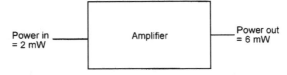

Power in = 2 mW — Amplifier — Power out = 6 mW

Insert the values:

$$\text{gain} = 10\log\left(\frac{6 \times 10^{-3}}{2 \times 10^{-3}}\right) \text{dB}$$

Simplify by dividing out the figures in the brackets. This gives:

gain = 10log3 dB

Take the log of 3:

gain = 10 × 0.477 dB

Simplify by multiplying:

gain = +4.77 dB

So this amplifier could be said to have a gain of 4.77 dB.

An attenuator, as shown in Figure 5.2, has less output power than input power:

$$\text{gain} = 10\log\left(\frac{\text{power}_{out}}{\text{power}_{in}}\right) \text{dB}$$

Figure 5.2

Find the loss in decibels

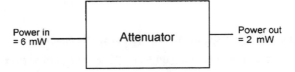

Power in = 6 mW — Attenuator — Power out = 2 mW

Keep the same formula for gains *and* losses. It makes life simpler. To prove it, we're just going to follow the same steps as in the last example.

Insert the values:

$$\text{gain} = 10\log\left(\frac{2 \times 10^{-3}}{6 \times 10^{-3}}\right) \text{dB}$$

Simplify by dividing out the contents of the brackets:

41

gain = 10log0.03334 dB

Taking the log we get:

gain = 10 × (–0.477) dB

Multiplying out we have:

gain = –4.77 dB.

Note

This attenuator can be described in two ways and it is vital that we do not get them confused.

The mathematical result was –4.77 dB. If we were to ask someone what the result was, they may well answer, 'The attenuator has a loss of 4.77 dB', or they may reply, 'It has a gain of minus 4.77 dB'.

In the first answer the fact that they mentioned 'loss' will alert us to the fact that an attenuation or loss has occurred. In the second example, the minus sign serves the same purpose.

We must be very careful not to fall into the double negative trap. It is best to avoid saying that the system has an overall loss of –4.77 dB. The 'loss' and the 'minus' could leave the sentence open to differing interpretations.

To summarize

If the number of decibels is negative, the result is a loss of power or an attenuation. If the result is positive, it indicates a gain or an amplification:

> ▷ + = gain/amplification
> ▷ – = loss/attenuation

The big advantage in using decibels is when a circuit consists of several gains and losses as will happen in real situations.

Decibels in a real circuit — what is the output power of the circuit shown in Figure 5.3?

Method

1   Express each change of power level in decibels.

An amplifier with a gain of 12 dB is represented by +12 dB

A loss of 16 dB is shown as –16 dB

Figure 5.3
What is the output
power in watts?

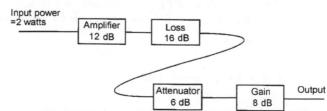

An attenuation of 6 dB is shown as –6 dB

A gain of 8 dB is shown as +8dB

Add all the decibels to give an overall result:

$$(+12) + (-16) + (-6) + (8)$$

Taking away the brackets:

$$+12 - 16 - 6 + 8 = -2 \text{ dB}.$$

The result is an overall loss of 2 dB so the circuit could be simplified to that shown in Figure 5.4.

Figure 5.4

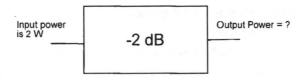

Input power is 2 W — **-2 dB** — Output Power = ?

2   **Looking at the formula for decibels:**

$$\text{gain} = 10\log\left(\frac{\text{power}_{\text{out}}}{\text{power}_{\text{in}}}\right) \text{ dB}$$

and knowing the result to be -2 dB and the input power to be 2 W, we can write:

$$-2 = 10\log\left(\frac{\text{power}_{\text{out}}}{2}\right) \text{ dB}$$

In this formula we know everything except the power out so we can transpose to find it.

It isn't a very obvious transposition so we will work through it one step at a time.

3   **Dividing both sides by 10 we have:**

$$\frac{-2}{10} = \log\left(\frac{\text{power}_{\text{out}}}{2}\right) \text{ dB}$$

or:

$$-0.2 = \log\left(\frac{\text{power}_{\text{out}}}{2}\right) \text{ dB}$$

4   **To undo the effect of the 'log' we must 'antilog' it**

This is done by raising both sides to a power of ten.

On the right hand side, this has the effect of removing the 'log' from the equation as shown:

$$10^{-0.2} = \frac{\text{power}_{\text{out}}}{2}$$

43

5   **Simplify the left hand side:**

$$10^{-0.2} = 0.63$$

So:

$$0.63 = \frac{power_{out}}{2}$$

Calculator note: the left hand side can be simplified by punching in –0.2 and hitting the 10^x or the inv log button. The –0.2 is often entered as 0.2 followed by the –/+ key.

6   **Multiplying both sides by 2 gives:**

$$1.26 = power\ out$$

and since we have been working in watts the output power is 1.26 watts.

Notice how we made life easy by simplifying the circuit to obtain a single overall figure in decibels before the conversion. Masochists, on the other hand, may enjoy the difficult approach of calculating the power out of the first amplifier in watts, then repeating the whole procedure to find the power out of the next section and so on through the circuit.

A summary of converting decibels to a ratio of two powers

Method

1   **Divide both sides by 10.**

2   **Find the antilog of each side.**

3   **Transpose to find the wanted term.**

Using a decibel as a power level

The essential point about a decibel is that it is used to express the ratio between two powers.

In the formula:

$$gain = 10\log\left(\frac{power_{out}}{power_{in}}\right) dB$$

we have two power levels mentioned, the *output power* and the *input power.*

If we wish to use decibels as a measurement of power, we have to get around the ratio problem by assuming a value for the input power.

So the formula is easily changed from:

$$gain = 10\log\left(\frac{power_{out}}{power_{in}}\right) dB$$

to:

$$\text{power level in decibels} = 10\log\left(\frac{\text{power level}}{\text{assumed power level}}\right) \text{dB}$$

The assumed power is usually 1 mW so the formula becomes:

$$\text{power level in decibels} = 10\log\left(\frac{\text{power level}}{1 \text{ mW}}\right) \text{dB}$$

For this to be of use, two points are important:

1. There must be a standard power level which is assumed and understood by everyone. In fiber optics, as with other branches of electronics, we use 1 mW.

2. We must indicate that we are now referring to a power level. This is done by changing the symbol to dBm, where dBm means *decibels relative to 1 mW*

Converting power to dBm

Example: express 5 Watts as a power level in decibels.

Method

1   Start with the formula:

$$\text{power level in decibels} = 10\log\left(\frac{\text{power level}}{1 \text{ mW}}\right) \text{dBm}$$

2   Put in the 5 watts:

$$\text{power level in decibels} = 10\log\left(\frac{5}{1 \times 10^{-3}}\right) \text{dBm}$$

3   Simplify by dividing out the bracket:

$$\text{power level in decibels} = 10\log(5 \times 10^{3}) \text{dBm}$$

4   Take the log of 5×10^{3}:

$$\text{power level in decibels} = 10 \times 3.699 \text{ dBm}$$

5   Multiply out:

$$\text{power level in decibels} = 36.99 \text{ dBm}$$

so:

5 watts = 36.99 dBm

Example: Converting dBm to a power level

A light source for a fiber optic system has an output power quoted as −14 dBm. Express this power in watts.

Method

1   **Always start with the formula:**

$$\text{power level in decibels} = 10\log\left(\frac{\text{power level}}{1 \text{ mW}}\right) \text{ dBm}$$

2   **Put in the figures that we know:**

$$-14 = 10\log\left(\frac{\text{power level}}{1 \times 10^{-3}}\right) \text{ dBm}$$

3   **Divide both sides by 10:**

$$-1.4 = \log\left(\frac{\text{power level}}{1 \times 10^{-3}}\right) \text{ dBm}$$

4   **Find the antilog of both sides:**

$$0.0398 = \frac{\text{power level}}{1 \times 10^{-3}} \text{ dBm}$$

5   **Multiply both sides by 1×10^{-3}:**

$$0.0398 \times 1 \times 10^{-3}$$

or:

$$39.8 \times 10^{-6} = \text{power level}$$

So a power of –14 dBm is the same as 39.8 μW.

Decibels used in a system design

Work in decibels for as long as possible. Converting backwards and forwards between decibels and watts is no fun and it is easy to make mistakes.

The nice thing is that dBm and dB are completely compatible — we can just add them up around the circuit to get a final result.

Method of solving the problem in Figure 5.5.

1   **Add the decibels:**

$$+16 \text{ dBm} + 3 \text{ dB} + (-8 \text{ dB}) -2 \text{ dB}$$

$$= +16 + 3 - 8 - 2 \text{ dBm}$$

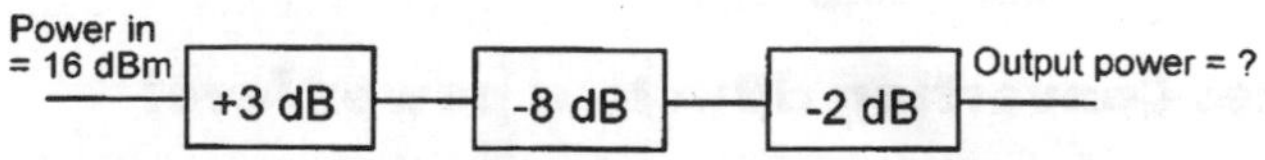

Figure 5.5

What is the output power in watts?

Power in = 16 dBm → +3 dB — -8 dB — -2 dB → Output power = ?

$$= +9$$

As the input power is quoted in dBm, the output power is +9 dBm.

To find the output power in watts, we must convert our answer into a power level as we did in the previous example.

2   **Start with the formula:**

$$\text{power level in decibels} = 10\log\!\left(\frac{\text{power level}}{1\ \text{mW}}\right) \text{dBm}$$

Put in the figures that we know:

$$9 = 10\log\!\left(\frac{\text{power level}}{1 \times 10^{-3}}\right) \text{dBm}$$

3   **Divide both sides by 10:**

$$0.9 = \log\!\left(\frac{\text{power level}}{1 \times 10^{-3}}\right) \text{dBm}$$

4   **Take the antilog of both sides:**

$$7.943 = \frac{\text{power level}}{1 \times 10^{-3}}$$

5   **Multiply both sides by 1×10^{-3}:**

$$7.943 \times 10^{-3} = \text{power level}$$

So a power of +9 dBm is the same as 7.943 mW.

Some dB values that are worth remembering:

> ▷   −3 dB = half power
> ▷   3 dB = a doubling of power
> ▷   10 dB = a tenfold increase
> ▷   −10 dB = a tenth power

Power loss on an optic fiber

Light power on an optic fiber is lost during transmission either by leakage or due to lack of clarity of the material.

The loss is expressed in decibels per kilometer and is written as dBkm^{-1}.

For silica glass fibers we are looking at values around 3 dBkm^{-1} for the fibers used for medium range transmissions. This corresponds to about half the power being last for each kilometer of travel. For long distance telecommunication fiber the figures are typically 0.3 dBkm^{-1} giving losses of only 7% per kilometer.

If a kilometer has a loss of 3 dB, then 2 km will have a total loss of 2 × 3 = 6 dB

It is, after all, just the same as having two attenuators connected in series. So, to obtain the total loss of a fiber, we simply multiply the loss specification in dBkm^{-1} by the length of the fiber (measured in kilometers of course).

Example

What is the output power, in watts, in the circuit shown in Figure 5.6?

Method

Figure 5.6

What is the output power in watts?

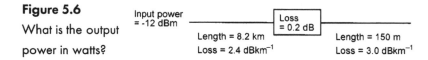

Input power = -12 dBm

Loss = 0.2 dB

Length = 8.2 km
Loss = 2.4 dBkm^{-1}

Length = 150 m
Loss = 3.0 dBkm^{-1}

1   Loss in a length of fiber = loss per km × length in km

So in the first length we have a loss of:

2.4 × 8.2 = 19.68 dB

and in the second case:

3.0 × 0.15 = 0.45 dB

(remembering to convert the 150 m to 0.15 km).

2   The circuit can now be simplified to that shown in Figure 5.7.

3   This gives a total of – 12 – 19.68 – 0.2 – 0.45 = –32.33 dBm

4   Convert this output power to watts by our previous method, giving a result of 0.58 μW.

Figure 5.7

A simplified version

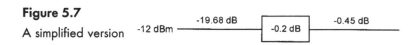

-12 dBm

-19.68 dB

-0.2 dB

-0.45 dB

Quiz time 5

In each case choose the best option:

1 **A power level of 50 µW could be expressed as:**

(a) 1.69 dBm

(b) -4.3 dBm

(c) 1 dBm

(d) -13 dBm

2 **The length of the link shown in Figure 5.8 is:**

(a) 1 km

(b) 3 km

(c) 5 km

(d) 7 km

Figure 5.8

All fiber used has a loss of 3.0 dBkm^{-1}

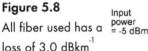

Input power = -5 dBm | Loss = 2 dB | Link A | Loss = 4 dB | 2 km of fiber | Output power = -26 dBm

All fiber used has a loss of 3.0 dBkm^{-1}

3 **If a power of 0.25 mW is launched into a fiber system with an overall loss of 15 dB the output power would be:**

(a) 250 µW

(b) 31.6 µW

(c) 7.9 µW

(d) 15 dBm

4 **A system having an input power of 2 mW and an output power of 0.8 mW has a loss of:**

(a) 2.98 dBm

(b) 3.98 dB

(c) 3.98 µW

(d) 1.98 mW

5 **An output of −10 dB means that the power has been:**

(a) halved in value

(b) increased by a factor of 10

(c) reduced by a factor of 10

(d) doubled

6

Losses in optic fibers

In the last chapter we saw that losses of 3 dB (50% of the power) are often incurred over a single kilometer. The exact figures depend on the fiber in use and we will be having a closer look at this in Chapter 7. So, for the moment, where does the light go — and why?

Basically, there are just two ways of losing light. Either the fiber is not clear enough or the light is being diverted in the wrong direction.

We will start with the first problem.

Absorption

Any impurities that remain in the fiber after manufacture will block some of the light energy. The worst culprits are hydroxyl ions and traces of metals.

The hydroxyl ions are actually the form of water which caused the large losses at 1380 nm that we saw in Figure 3.2. In a similar way, metallic traces can cause absorption of energy at their own particular wavelengths. These small absorption peaks are also visible.

In both cases, the answer is to ensure that the glass is not contaminated at the time of manufacture and the impurities are reduced as far as possible. We are aiming at maximum levels of 1 part in 10^9 for water and 1 part in 10^{10} for the metallic traces.

Now for the second reason, the diversion of the light.

Rayleigh scatter

This is the scattering of light due to small localized changes in the refractive index of the core and the cladding material. The changes are indeed very localized. We are looking at dimensions which are less than the wavelength of the light.

There are two causes, both problems within the manufacturing processes.

The first is the inevitable slight fluctuations in the 'mix' of the ingredients. These random changes are impossible to completely eliminate. It is a bit like making a currant bun and hoping to stir it long enough to get all the currants equally spaced.

The other cause is slight changes in the density as the silica cools and solidifies.

One such discontinuity is illustrated in Figure 6.1 and results in light being scattered in all directions. All the light that now finds itself with an angle of incidence less than the critical angle can escape from the core and is lost. However, much of the light misses the discontinuity because it is so small. The scale size is shown at the bottom.

Figure 6.1
The light is scattered in all directions

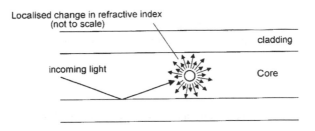

This is the scale size ⟶·

The amount of scatter depends on the size of the discontinuity compared with the wavelength of the light so the shortest wavelength, or highest frequency, suffers most scattering. This accounts for the blue sky and the red of the sunset. The high frequency end of the visible spectrum is the blue light and this is scattered more than the red light when sunlight hits the atmosphere. The sky is only actually illuminated by the scattered light. So when we look up, we see the blue scattered light, and the sky appears blue. The moon has no atmosphere, no scattering, and hence a black sky. At sunset, we look towards the sun and see the less scattered light which is closer to the sun. This light is the lower frequency red light.

Fresnel reflection

On a point of pronunciation, the *s* in Fresnel is silent.

When a ray of light strikes a change of refractive index and is approaching at an angle close to the normal, most of the light passes straight through as we saw in a previous chapter.

Most of the light but not all. A very small proportion is reflected back off the boundary. We have seen this effect with normal window glass. Looking at a clean window we can see two images. We can see the scene in front of us and we can also see a feint reflection of what is behind us. Light therefore is passing through the window and is also being reflected off the surface.

We are most concerned about this loss when considering the light leaving the end of the fiber as shown in Figure 6.2. At this point, we have a sudden transition between the refractive index of the core and that of the surrounding air. The effect happens in the other direction as well. The same small proportion of light attempting to enter the fiber is reflected out again as in Figure 6.3.

The actual proportion of the light is determined by the amount by which the refractive index changes at the boundary and is given by the formula:

$$\text{reflected power} = \left(\frac{n_1 - n_2}{n_1 + n_2}\right)^2$$

To see how bad it can get, let's take a worst case situation — a core of refractive index 1.5 and the air at 1.0.

Method

Start by inserting the figures in the above equation:

$$\text{reflected power} = \left(\frac{1.5 - 1.0}{1.5 + 1.0}\right)^2$$

Simplify the top and bottom terms inside the brackets:

$$\text{reflected power} = \left(\frac{0.5}{2.5}\right)^2$$

Divide out the terms in the bracket:

$$\text{reflected power} = (0.2)^2 = 0.04 = 4\%$$

So, 96% of the incident light power penetrates the boundary and the other 4% is reflected. This reflected power represents a loss of 0.177 dB.

It may be worth mentioning in passing, that if we try to squirt light from one fiber into another, we suffer this 0.177 dB loss once as the light leaves the first fiber and then again as the light attempts to enter the other fiber.

Remember that these figures are worst case. We get up to all sorts of tricks to improve matters as we shall see when we look at ways of connecting lengths of optic fibers together.

Making use of Fresnel reflection

The return of the Fresnel reflection from the end of a fiber gives us a convenient and accurate method of measuring its length. Imagine a situation in which we have a drum of fiber optic cable marked 5 km. Does the drum actually contain 5 km? or 4.5 km? or is it in five separate lengths of 1 km? It is inconvenient, to say the least, to uncoil and measure all fiber as it is delivered.

The solution is to make use of Fresnel reflection that will occur from the far end. We send a short pulse of light along the fiber and wait for the reflection to

Figure 6.2

Fresnel reflection

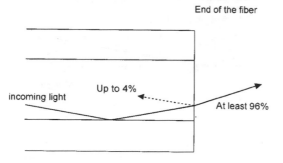

Figure 6.3

A Fresnel reflection
from each surface

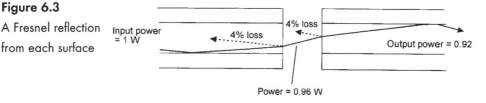

bounce back. Since we can calculate how fast the light is traveling and can measure the time interval, the length is easily established. This magic is performed for us by an instrument called an *optical time domain reflectometer* (OTDR), as discussed in Chapter 15.

Bending losses

Macrobends

A sharp bend in a fiber can cause significant losses as well as the possibility of mechanical failure.

It is easy to bend a short length of optic fiber to produce higher losses than a whole kilometer of fiber in normal use.

The ray shown in Figure 6.4 is safely outside of the critical angle and is therefore propagated correctly.

Remember that the normal is always at right angles to the surface of the core. Now, if the core bends, as in Figure 6.5, the normal will follow it and the ray will now find itself on the wrong side of the critical angle and will escape.

Tight bends are therefore to be avoided but how tight is tight?

The real answer to this is to consult the specification of the fiber optic cable in use as the manufacturer will consider the mechanical limitations as well as the bending losses. However a few general indications may not be out of place.

A bare fiber — and by this is meant just the core/cladding and the primary buffer — is safe if the radius of the bend is at least 50 mm. For a cable, which is the bare fiber plus the outer protective layers, make it about 10 x outside diameter or 50 mm whichever is the greater (Figure 6.6).

53

Figure 6.4

The usual situation

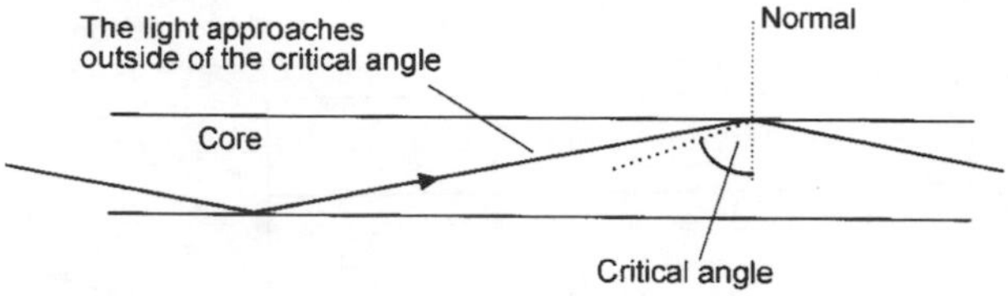

The light approaches
outside of the critical angle

Normal

Core

Critical angle

Figure 6.5

Sharp bends are
bad news

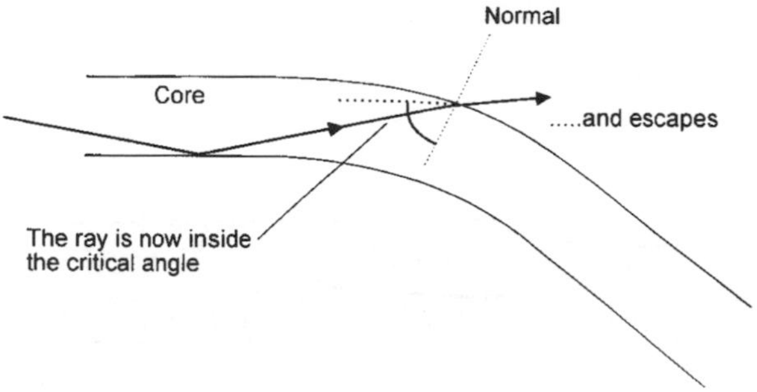

Normal

Core

.....and escapes

The ray is now inside
the critical angle

Figure 6.6

Minimum safe bend
radius — shown full
size

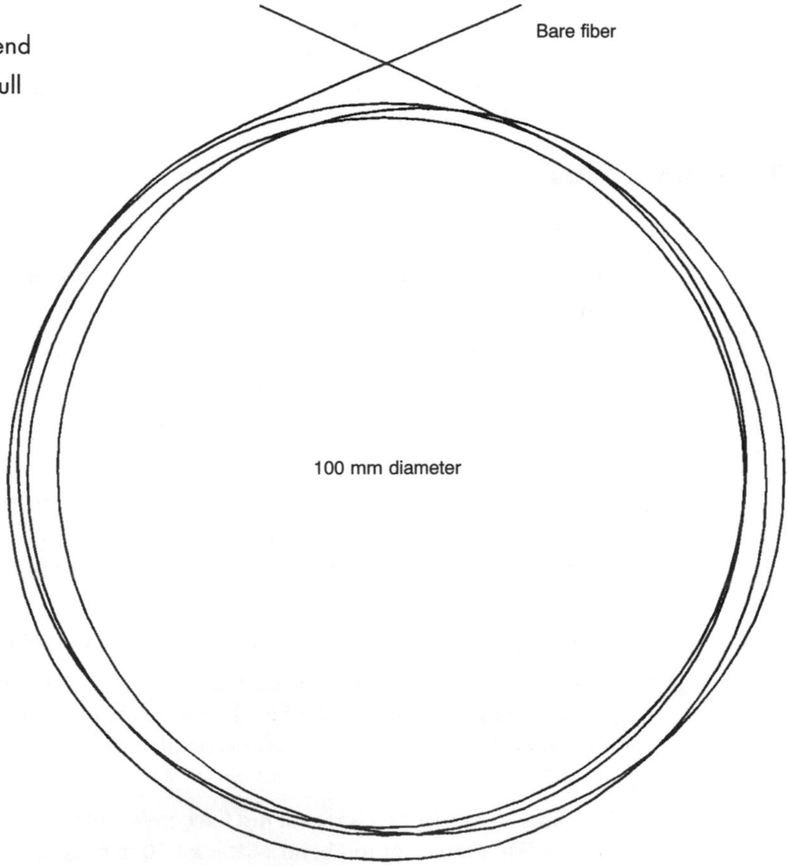

Bare fiber

100 mm diameter

The tighter the bend, the worse the losses. Shown full size, the results obtained with a single sample of bare fiber were shown in Figure 6.7, attached instruments indicated a loss of over 6 dB before it broke.

The problem of macrobend loss is largely in the hands of the installer.

Figure 6.7

Bends are shown full size – and may have caused damage to the fiber

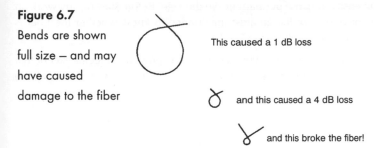

This caused a 1 dB loss

and this caused a 4 dB loss

and this broke the fiber!

Making use of bending losses

There are many uses of bending losses which are based on either the increase in the attenuation or on making use of the light which escapes from the optic fiber.

A fiber optic pressure sensor

This makes use of the increased attenuation experienced by the fiber as it bends.

A length of bare fiber is sandwiched between two serrated pieces of rubber or plastic matting. The fiber is straight and the light detector is 'on guard' at the far end (Figure 6.8).

Figure 6.8

The fiber is straight when no pressure is applied

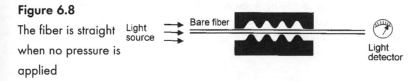

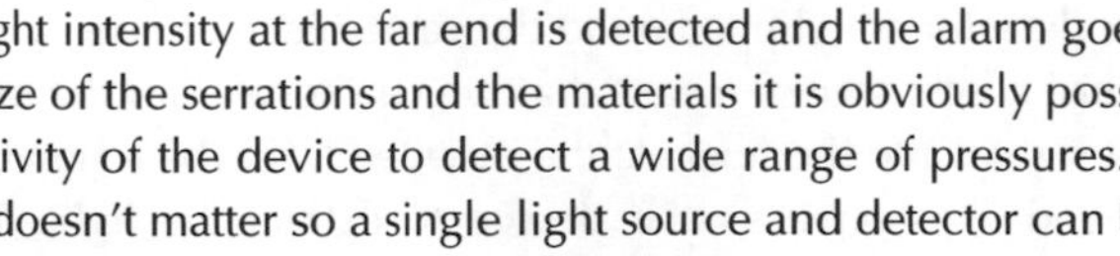

Light source

Bare fiber

Light detector

Someone steps on the mat (Figure 6.9). Bends are created and the reduction in the light intensity at the far end is detected and the alarm goes off. By changing the size of the serrations and the materials it is obviously possible to change the sensitivity of the device to detect a wide range of pressures. The length of the fiber doesn't matter so a single light source and detector can be used to monitor many pressure pads at the same time.

Figure 6.9

Pressure causes loss at the bends

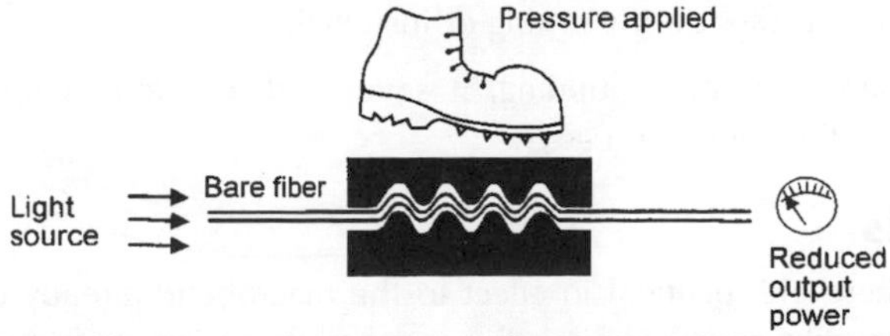

Pressure applied

Light source

Bare fiber

Reduced output power

Active fiber detector

This uses the escaping light.

Many fibers used in telecommunications carry invisible infrared light of sufficient intensity to cause permanent eye damage. Before starting any work on a fiber it is obviously of the greatest importance to know whether the fiber is carrying light and is therefore 'live' or 'active'.

It works on a principle which is very similar to the pressure detector except that it sniffs the light escaping. One example of such as device is shown in Figure 6.10.

Figure 6.10

Is the fiber in use?

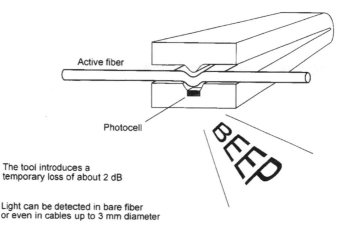

Active fiber

Photocell

The tool introduces a
temporary loss of about 2 dB

Light can be detected in bare fiber
or even in cables up to 3 mm diameter

The fiber is placed between the jaws of the tool and, as they close, a sharp bend is created and some of the light escapes. The escaping light is detected by a photocell and is used to switch on a warning light or beeper. Some versions will also detect the direction of the traffic on the fiber.

Talkset

As all light paths are reversible, it follows that a sharp bend can allow light to enter the fiber as well as to leave it. Two devices similar in outline to the active fiber detector are clipped on to the same piece of fiber. Using a microphone to control a light source means that we can speak into the microphone and put a light signal onto the fiber. At the other end, maybe a hundred kilometers away, a friend clips on a similar device to extract the light. The output from the photocell, after suitable amplification, is fed to a pair of headphones. Putting both a transmitter and a receiver at each end allows communication between the two sites during testing or installation.

Even in a small building, it saves endless running up and down the stairs shouting 'are you ready?'

Microbends

These are identical in effect to the macrobend already described but differ in size and cause. Their radius is equal to, or less than, the diameter of the bare fiber — very small indeed (Figure 6.11).

Figure 6.11

Differential

contraction can

cause microbends

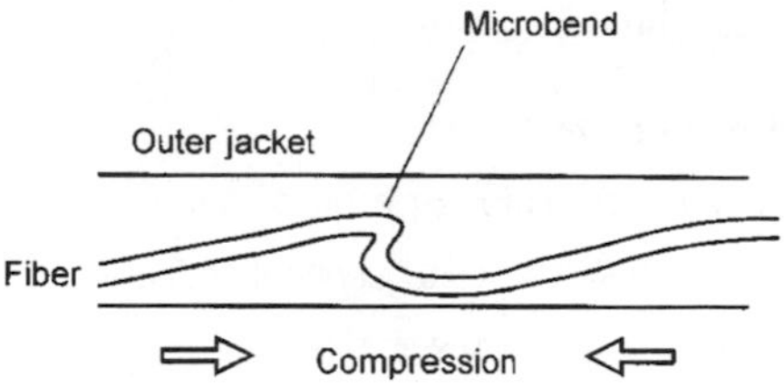

These are generally a manufacturing problem. A typical cause is differential expansion of the optic fiber and the outer layers. If the fiber gets too cold, the outer layers will shrink and get shorter. If the core/cladding shrinks at a slower rate, it is likely to kink and cause a microbend.

With careful choice of the fiber to be installed, these are less likely to be a problem than the bending losses caused during installation since fiber optic cables are readily available with a wide range of operating temperatures from -55 °C to +85 °C.

Quiz time 6

In each case, choose the best option.

1   An active fiber detector:

(a) is used to prevent accidental exposure to invisible light

(b) is a communication system used during installation

(c) detects movement of fiber in security systems

(d) can be used to weigh objects

2   If light leaves a material of refractive index 1.45 and crosses an abrupt boundary into a material of refractive index 1.0, the Fresnel loss would be:

(a) 0.346 dB

(b) 0.149 dB

(c) 1.613 dB

(d) 3 dB

3   Absorption loss is caused by:

(a) insufficient stirring of the ingredients during manufacture

(b) changes in the density of the fiber due to uneven rates of cooling

(c) microscopic cracks in the cladding which allow leakage of the vacuum in the core

(d) impurities in the fiber

4   Rayleigh scatter is most severe:

(a) in light with a short wavelength

(b) in blue light on the Earth and black light on the moon

(c) in low frequency light

(d) at sunset

5   Bending losses:

(a) always result in breakage of the fiber

(b) can be caused by microbends and macrobends

(c) are used to detect the length of fiber on a drum

(d) are caused by the difference in the operating temperature of the core compared with the cladding on active fibers

7

Dispersion and our attempts to prevent it

In the last chapter we looked at a few causes of power loss which tend to limit the useful transmission range. Unfortunately the mere fact that we can send light to the other end of the system is not enough, we have also to ensure that the data is still decipherable when it arrives.

We have met the effect in public address systems. As soon as the announcement ends, people turn to one another and say 'What did she say?' 'What was that all about?'. We can't make out what is being said. Hearing the sound is no problem, there is plenty of volume and increasing it further would not help and would probably make the situation worse.

The problem is that the sound is arriving by many separate paths of different lengths, and parts of each sound are arriving at different times. In most cases, we can handle this well and automatically compensate for any multiple reflections but in severe situations it simply overloads our brain's computing capability. Indeed we are so clever at handling this situation that we can use it to 'sense' the size of a room. When we step over the threshold the spurious echoes cease and we know that we are outside.

Optic fibers suffer from a very similar effect, called *dispersion*.

Dispersion

Imagine we launch two different rays of light into a fiber. Now, since both rays are traveling in material of the same refractive index they must be moving at the same speed. If we follow two different rays of light which have entered the core at the same time, we can see that one of them, Ray A in Figure 7.1, will travel a longer distance than the other, Ray B. The effect of this is to cause the pulse of light to spread out as it moves along the fiber — as the ray taking the shorter route overtakes the other.

This spreading effect is called dispersion (Figure 7.2).

Figure 7.1

Ray B will arive first

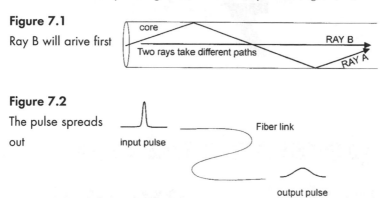

Figure 7.2

The pulse spreads

out

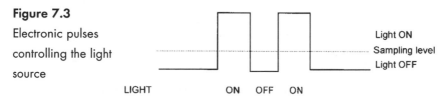

The effect on the data

The data can be corrupted by dispersion. If we send a sequence of ON OFF ON pulses, it would start its life as an electronic signal with nice sharp edges as in Figure 7.3.

Figure 7.3

Electronic pulses

controlling the light

source

These pulses are used to switch a light source, usually an LED or a laser and the resultant pulses of light are launched into the fiber.

Dispersion causes the pulses to spread out and eventually they will blend together and the information will be lost (Figure 7.4).

We could make this degree of dispersion acceptable by simply decreasing the transmission frequency and thus allowing larger gaps between the pulses.

This type of dispersion in called *intermodal dispersion*. At this point, we will take a brief detour to look at the idea of modes.

Modes

The light traveling down the fiber is a group of *electromagnetic* (EM) waves occupying a small band of frequencies within the electromagnetic spectrum, so it is a simplification to call it a ray of light. However, it is enormously helpful to do this, providing an easy concept, some framework to hang our ideas on. We do this all the time and it serves us well providing we are clear that it is only an

Figure 7.4

Dispersion has

caused the pulses

to merge

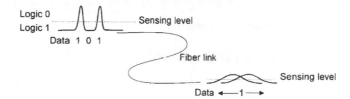

analogy. Magnetic fields are not really lines floating in space around a magnet, electrons are not really little black ball bearings flying round a red nucleus.

Light therefore, is propagated as an electromagnetic wave along the fiber. The two components, the electric field and the magnetic field form patterns across the fiber. These patterns are called *modes of transmission*. Modes means methods — hence methods of transmission. An optic fiber that carries more than one mode is called a *multimode* fiber (MM).

The number of modes is always a whole number.

In a given piece of fiber, there are only a set number of possible modes. This is because each mode is a pattern of electric and magnetic fields having a physical size. The dimensions of the core determine how many modes or patterns can exist in the core — the larger the core, the more modes.

The number of modes is always an integer, we cannot have incomplete field patterns. This is similar to transmission of motor vehicles along a road. As the road is made wider, it stays as a single lane road until it is large enough to accommodate an extra line of vehicles whereupon it suddenly jumps to a two lane road. We never come across a 1.15 lane road!

How many modes are there?

The number of modes is given (reasonably accurately) by the formula:

$$\text{Number of modes} = \frac{\left(\text{Diameter of core} \times \text{NA} \times \dfrac{\pi}{\lambda}\right)^2}{2}$$

where NA is numerical aperture of the fiber and λ is the wavelength of the light source.

Let's choose some likely figures as in Figure 7.5 and see the result.

Figure 7.5

How many modes

are in the core?

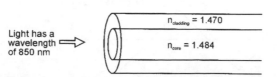

Core diameter = 50 μm

Light has a
wavelength
of 850 nm

$n_{cladding}$ = 1.470

n_{core} = 1.484

Method

Find the numerical aperture

$$NA = \sqrt{(1.484)^2 - (1.470)^2}$$

So:

NA = 0.203 (a typical result)

Insert the figures into the formula:

$$\frac{\left(50 \times 10^{-6} \times 0.203 \times \dfrac{\pi}{850 \times 10^{-9}}\right)^2}{2}$$

Tap it into a calculator and see what comes out:

number of modes = 703.66

The calculator gave 703.66 but we cannot have part of a mode so we have to round it down. Always round down as even 703.99 would not be large enough for 704 modes to exist.

Each of the 703 modes could be represented by a ray being propagated at its own characteristic angle. Every mode is therefore traveling at a different speed along the fiber and gives rise to the dispersion which we called intermodal dispersion.

How to overcome intermodal dispersion

We can approach the problem of intermodal dispersion in two ways. We could redesign the fiber to encourage the modes to travel at the same speed along the fiber or we can eliminate all the modes except one — it can hardly travel at a different speed to itself! The first strategy is called *graded index* optic fiber.

Graded index fiber

This design of fiber eliminates about 99% of intermodal dispersion. Not perfect — but a big improvement.

The essence of the problem is that the ray that arrives late has taken a longer route. We can compensate for this by making the ray that takes the longer route move faster. If the speed and distance of each route is carefully balanced, then all the rays can be made to arrive at the same time — hence no dispersion. Simple. At least in theory (Figure 7.6).

Figure 7.6

They arrive at (almost) the same time

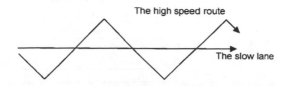

The high speed route

The slow lane

We often meet this while driving. Do we take the shorter route and creep through the city center? Or do we take the longer, faster route on the by-pass? So often we find it makes little difference, the extra distance is offset by the extra speed.

The speed that light travels in the core is determined by the refractive index.

The formula is:

$$\text{Speed of light in the material} \ = \ \frac{\text{speed of light in free space}}{\text{refractive index}}$$

The solution to our problem is to change the refractive index progressively from the center of the core to the outside. If the core center has the highest refractive index and the outer edge has the least, the ray will increase in speed as it moves away from the center.

The rate at which the refractive index changes is critical and is the result of intensive research. A parabolic profile is often employed but there are many others available in specialized fibers.

The GI (graded index) and the SI (step index) profiles are shown in Figure 7.7. Step index fibers are the basic type in which the core has a set value of refractive index and is surrounded by the cladding, with its lower value. This results in the characteristic step in the value of the refractive index as we move from the core to the cladding.

Figure 7.7

Two types of fiber. Top is step index fiber. Bottom is graded index fiber

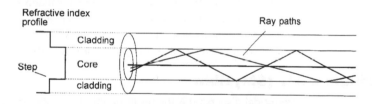

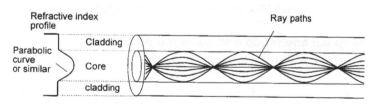

We can see that, in the GI fiber, the rays each follow a curved path. This is one of the results of the change in refractive index as we move away from the center of the core.

We can consider the core to be made of a whole series of discrete changes in refractive index as shown in Figure 7.8. At each boundary there is a change in refractive index and the light ray is refracted slightly. Every time the ray is refracted the angle of incidence increases. Eventually the ray will approach a layer at an angle greater than the critical angle and reflection occurs.

It is now approaching the core and as it passes through the layers of increasing

Figure 7.8

The ray is refracted
slight at each
boundary

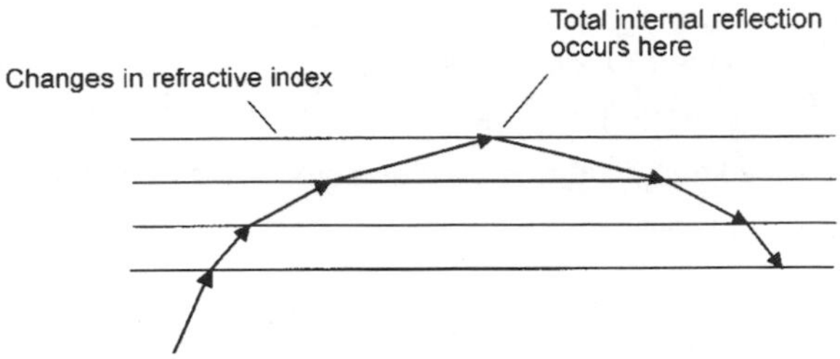

Changes in refractive index

Total internal reflection
occurs here

refractive indices the curvature will again increase. This process will be repeated as it crosses the fiber and the light will be made to follow a curved path along the core. Notice how some rays (modes) are restricted to the center (low speed) area of the core and some extend further out into the faster regions.

It is important to appreciate that, in a real GI fiber, change in the refractive index is smooth and continuous. It is not really arranged in layers as is suggested by the diagram. The result is that the light suffers an infinite number of small refractions and has the effect of making the light bend in the smooth curves we saw in Figure 7.7, rather than discrete steps at each layer.

The number of modes in a graded index fiber is half that found in a step index fiber so in the mode formula the bottom line should read the figure 4, rather than the 2 shown. The practical result of this is that a step index fiber will hold twice the number of modes and hence accept twice the input power from a light source than would a graded index fiber. However, the dispersion advantage of the graded index still makes a graded index desirable for multimode fibers.

Single mode (SM) fiber

Intermodal dispersion is the result of different modes (rays) traveling at different speeds. The easy way to avoid this is to have only one mode.

How to get one mode and solve the problem

If we have another look at the formula for the number of modes:

$$\text{Number of modes} = \frac{\left(\text{Diameter of core} \times \text{NA} \times \frac{\pi}{\lambda}\right)^2}{2}$$

we can see that we could decrease the number of modes by increasing the wavelength of the light. However this alone cannot result in reducing the number of modes to 1. Changing from the 850 nm window to the 1550 nm window will only reduce the number of modes by a factor of 3 or 4 which is not enough on its own. Similarly, a change in the numerical aperture can help but it only makes a marginal improvement.

We are left with the core diameter. The smaller the core, the fewer the modes. When the core is reduced sufficiently the number of modes can be reduced to just one. The core size of this SM or single mode fiber is between 5 μm and

10 μm. Figure 7.9 shows a MM and a SM fiber drawn to scale. The difference in the core size is clearly visible.

Figure 7.9

It's easy to see the difference

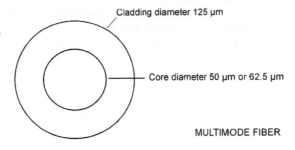

Cladding diameter 125 μm

Core diameter 50 μm or 62.5 μm

MULTIMODE FIBER

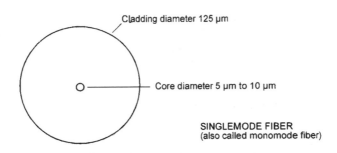

Cladding diameter 125 μm

Core diameter 5 μm to 10 μm

SINGLEMODE FIBER
(also called monomode fiber)

Intramodal (or chromatic) dispersion

Unfortunately intermodal dispersion is not the only cause of dispersion.

We know that light of different wavelengths is refracted by differing amounts. This allows rain drops to split sunlight into the colors of the rainbow. We are really saying that the refractive index, and hence the speed of the light, is determined to some extent by its wavelength.

A common fallacy is that a laser produces light of a single wavelength. In fact it produces a range of wavelengths even though it is far fewer than is produced by the LED, the alternative light source (Figure 7.10).

Figure 7.10

The laser has a narrow spectral width

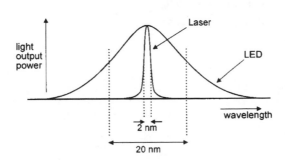

light output power

Laser

LED

wavelength

2 nm

20 nm

This is unfortunate as each component wavelength travels at a slightly different speed in the fiber. This causes the light pulse to spread out as it travels along the fiber — and hence causes dispersion. The effect is called *chromatic dispersion*.

Actually, chromatic dispersion is the combined effect of two other dispersions — *material dispersion* and *waveguide dispersion*. Both result in a change in transmission speed, the first is due to the atomic structure of the material and the second is due to the propagation characteristics of the fiber. Any further investigation will not help us at the moment and is not pursued further.

One interesting feature of chromatic dispersion is shown in Figure 7.11. The value of the resulting dispersion is not constant and passes through an area of zero dispersion. This cannot be used to eliminate dispersion altogether because the zero point only occurs at a single wavelength, and even a laser produces a range of wavelengths within its spectrum.

Figure 7.11

The effect of

wavelength

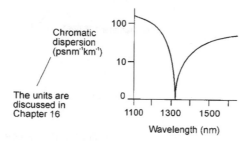

Chromatic
dispersion
(psnm⁻¹km⁻¹)

The units are
discussed in
Chapter 16

By fiddling about with the dimensions of the core and the constituents of the fiber, we can adjust the wavelength of the minimum dispersion point. This is called *optimizing* the fiber for a particular wavelength or window.

Be careful not to muddle intermodal dispersion with intramodal dispersion.

> ➤ *inter* means between. *Intermodal* — between modes

> ➤ *intra* means within. *Intramodal* — within a single mode

The alternative name of chromatic, to do with color or frequency, is less confusing.

Although chromatic dispersion is generally discussed in terms of a single mode fiber it does still occur in multimode fiber but the effect is generally swamped by the intermodal dispersion.

Dispersion is summarized in Figure 7.12.

Figure 7.12

Dispersion – a

summary

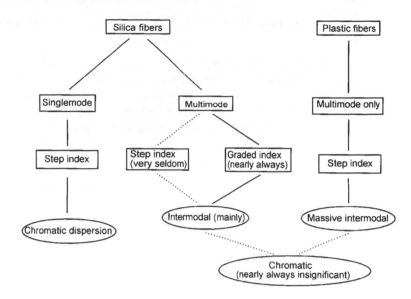

Quiz time 7

In each case, choose the best option.

1   Dispersion:

(a) causes the core to spread out and get wider as the pulse is transmitted along the fiber

(b) results in the wavelength of the light increasing along the fiber

(c) is the lengthening of light pulses as they travel down the fiber

(d) cannot occur with a laser light source

2   An SI MM fiber has a core of 62.5 μm diameter and a numerical aperture of 0.2424. The number of modes that would occur using a light of wavelength of 865 nm would be:

(a) 1

(b) 378

(c) 1513.78

(d) 1513

3   Intramodal dispersion:

(a) only occurs in multimode fiber

(b) is also called chromatic dispersion

(c) does not occur in multimode fiber

(d) could not occur in an all plastic fiber

4   If the wavelength of the transmitted light were to be decreased, the number of modes would :

(a) increase

(b) decrease

(c) remain the same

(d) halve in a graded index fiber

5   **The refractive index of a GI fiber:**

(a) is at its highest value at the center of the core

(b) is usually higher in the cladding than in the core

(c) increases as we move away from the center of the core

(d) has a value of 4 instead of the 2 common in step index fibers

8

Real cables

The optic fiber manufacturers provide the primary coated optic fiber. At this stage, the fiber is easily broken between finger and thumb and is typically only 250 μm in diameter.

The cable manufacturers then enclose the optical fiber (or fibers) in a protective sleeve or sleeves. These sleeves are often referred to as *jackets,* or *buffers* - the exact terminology differs a little according to manufacturer. Once enclosed, the assembly is then referred to as a *cable.*

The cost of installing the cable on site is much greater than the cost of the optic fiber alone, so it makes good sense to provide plenty of spare system capacity to allow for failures or expansion in the traffic to be carried. Nowadays, traffic that decreases in volume is a rare event indeed.

The degree of protection depends on the conditions under which the cable is to operate. Some cables will live a luxurious life, warm, dry and undisturbed asleep in a duct in an air-conditioned office while others are outside in the real world. These may well be submerged in water or solvents, attacked by rodents, at sub-zero temperatures or being crushed by earthmovers on a construction site.

Strength members

If we pull a length of fiber through a duct, the outer cover would stretch and the pulling load would be taken by the glass fiber, which would break. To prevent this, the cable is reinforced by adding strength members. These are strong low stretch materials designed to take the strain. The strength members used are usually fibers of Kevlar (® Dupont Inc) for light duty cable and fiber glass rods or steel for heavier duty cable. Kevlar is a very fine, yellow, silky fiber which is, weight for weight, about four times stronger than steel. It resists crushing and being pierced which makes it popular for another of its applications — bullet-proof clothing. Unfortunately, It stretches too much to replace steel altogether.

Basic choice of cable design

There are two distinctly different methods used to protect the optic fibers. They are referred to as *loose tube* and *tight buffer* designs. We will have a look at these in a moment. There is a tendency for tight buffer cables to find employment within buildings and loose tube designs to be used externally. This is not a 'golden rule', just a bias in this direction.

Loose tube construction

A hollow polymer tube surrounds the optic fiber as can be seen in Figure 8.1. The internal diameter of this tube is much greater than the diameter of the optic fibers that simply lie inside it. There is room for more than one fiber and as many as twelve optic fibers can run through the same tube. Note, in catalogs these possibilities are often referred to as *singlefiber* and *multifiber*. Be careful not to misread these as singlemode and multimode.

Figure 8.1

Loose tube

construction

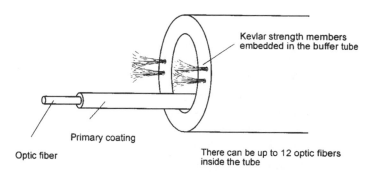

Kevlar strength members
embedded in the buffer tube

Primary coating

Optic fiber

There can be up to 12 optic fibers
inside the tube

The main feature is that the optic fiber is thus free to move about as it wishes. The benefit of this is that its natural springiness allows the optic fiber to take the path of least strain. Figure 8.2 shows how the fiber is able to take the route with the largest radius of curvature and help to protect itself from bending losses.

Under normal, non stress, conditions the optic fiber(s) tend to snake lazily about inside the buffer tube and this results in the optic fiber itself being slightly (about 1%) longer than the buffer tube. This has the further advantage that the cable can be stretched by about 1% during installation without stressing the optic fiber (Figure 8.3).

Tight buffer construction

In this case, a jacket is fitted snugly around the optic fiber in the same way that electrical cables are coated in plastic. This provides protection while allowing flexibility. This form of construction (shown in Figure 8.4) is normally, but again not exclusively, used for indoor installations. Tight jacketed cables come in a variety of forms to suit the installation requirements. A small selection is shown in Figure 8.5.

Figure 8.2
Some benefits of
loose tube
construction

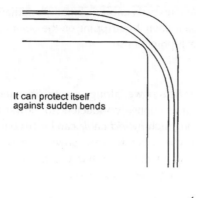

It can protect itself
against sudden bends

and the occasional crush

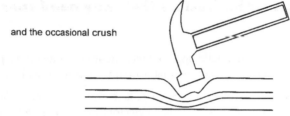

Figure 8.3
The optic fiber is up
to 1% longer than
the tube

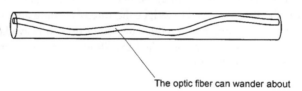

The optic fiber can wander about

Figure 8.4
Tight buffer
construction

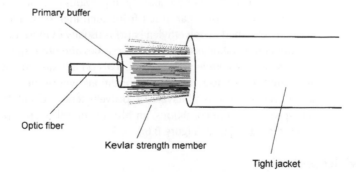

Primary buffer

Optic fiber

Kevlar strength member

Tight jacket

Plastic optic fiber cable is always tight buffered.

There are two options regarding the strength members. They can be placed immediately under the outer jacket or it can be added around each individual optic fiber within their own jacket, sometimes referred to as a *sub-jacket*.

Breakout cable

The advantage of the latter approach is that the outer jacket can be stripped off leaving us with individual cables, each with its own strength members. This type is called *breakout cable.* The main cable is installed, perhaps as far as an office space via ceiling ducts and then, by stripping off the outer jacket, the individual cables can be fed to each point of use.

Hybrid cable

We cannot transmit electrical power along an optic fiber and so, in some cases, it is necessary to include copper conductors to power repeaters or other instruments at the far end. Such *hybrid cable* can be custom built and may well include many different types of optic fiber, power cables of different voltage, copper coaxial cables and anything else that you feel like. The resulting cable can be bulky, heavy and very, very expensive.

Cable design - other factors that may need considering

Fire precautions

For use within buildings, the outer sheath is usually PVC (polyvinyl chloride) — the same plastic coating that is used for electrical wiring.

The PVC is often treated to reduce the fire hazards. As we know from the publicity surrounding smoke detectors in buildings, toxic fumes and smoke are an even greater hazard than the fierce burning of the PVC. To reduce the fire risk, the PVC is treated to reduce the rate of burning and is referred to as flame retardant. In addition, to reduce the other risks, it is available with low smoke and gas emission. This is called LSF (low smoke and fumes) and LSOH (low smoke, zero halogen).

Moisture

Unfortunately, PVC is not completely impervious to water and, as we have seen previously, large losses can result from such ingress. Cable for external use is normally sheathed in polyethylene and is usually of loose tube construction. An additional precaution is to fill the loose tubes and other spaces within the cable with a silicon or petroleum based gel. This gel is waterproof and thixotropic and flows into any spaces. In severe conditions where the cable is permanently submerged even several layers of polyethylene are not found to be totally waterproof. In these conditions, an aluminum foil wrap is placed immediately under the outer jacket (Figure 8.6).

Ultra violet protection

Sunlight and other sources of UV light tend to degrade many plastics, including standard polyethylene, in the course of time. In most cases, making the outer sheath from UV stabilized polyethylene is sufficient but for extreme situations a metal sheathed cable is used.

Hydrocarbons

Hydrocarbons are organic compounds of hydrogen and carbon and include solvents, oils, benzene, methane and petroleum. Polyethylene provides a satisfactory means of combating most hydrocarbons. For the ultimate in

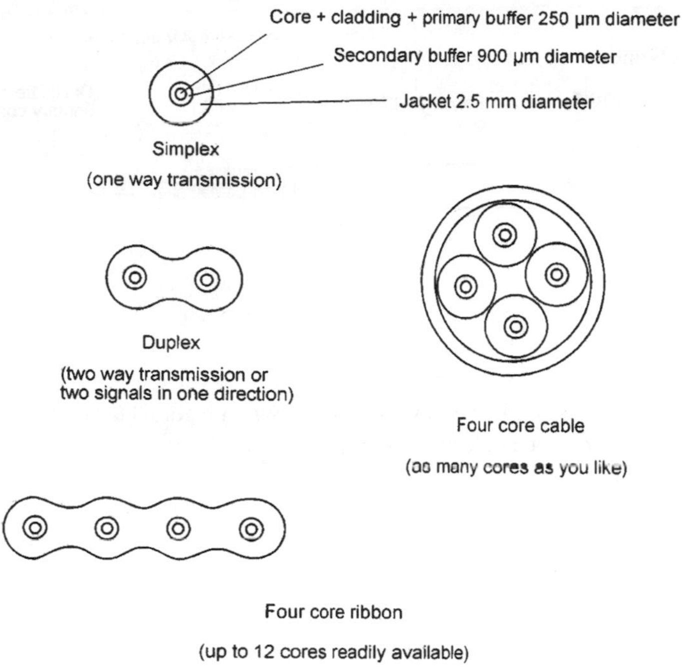

Figure 8.5
Tight jacketed cables (strength members omitted for clarity)

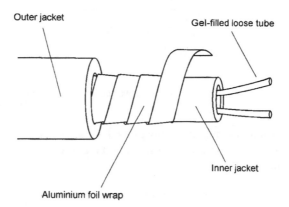

Core + cladding + primary buffer 250 µm diameter

Secondary buffer 900 µm diameter

Jacket 2.5 mm diameter

Simplex
(one way transmission)

Duplex
(two way transmission or two signals in one direction)

Four core cable
(as many cores as you like)

Four core ribbon
(up to 12 cores readily available)

Figure 8.6
A waterproof cable (strength members omitted for clarity)

Outer jacket

Gel-filled loose tube

Inner jacket

Aluminium foil wrap

protection against these agents, a lead inner sheath can be added under the polyethylene outer layer.

Mechanical damage

Mechanical support and strength members are not the same even though, in some cases, both functions can be performed by the same elements. Strength members are present to prevent the cable being excessively stretched but mechanical damage can involve other forms of abuse such as being crushed or cut.

In light duty cables as used in buildings, both functions can be achieved by a layer of Kevlar or similar material placed under the outer jacket (Figure 8.7).

We may also provide some localized protection such as when the fiber is fed

73

Figure 8.7

Light duty indoor
distribution cable

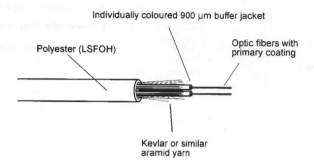

Individually coloured 900 μm buffer jacket

Polyester (LSFOH)

Optic fibers with
primary coating

Kevlar or similar
aramid yarn

under a carpet, for instance, it will be walked on and crushed by furniture being moved. A protective strip is shown in Figure 8.8 and is very similar to those used to protect electrical wiring.

Figure 8.8

A form of localized
protection

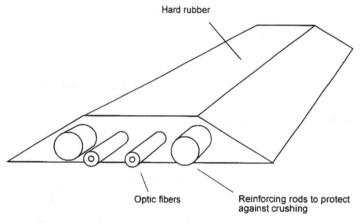

Hard rubber

Optic fibers

Reinforcing rods to protect
against crushing

In situations beyond the ability of Kevlar, the normal form of protection is to add a layer, or several layers, of metallic tape or galvanized wire under the outer sheath as in Figure 8.9.

Figure 8.9

An external direct
burial cable

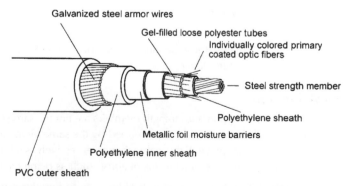

Galvanized steel armor wires

Gel-filled loose polyester tubes

Individually colored primary
coated optic fibers

Steel strength member

Polyethylene sheath

Metallic foil moisture barriers

Polyethylene inner sheath

PVC outer sheath

The cable may have to survive crushing forces from surface traffic or earth movement or perhaps by being picked up by a mechanical digger. It may be laid across a river or indeed an ocean. In the latter case, the cable is initially buried but for most of its distance it is simply dropped onto the ocean floor to take its chance. The enormous cost of location, recovery and repair make extra mechanical protection a very good investment. It is, of course, a matter of judgment as to the degree of protection that a particular cable requires. Adding further protection has the inevitable result of increasing the size and weight which will mean that the strength members would need upgrading to prevent damage during installation.

It is worth mentioning that the galvanized wires, added for additional crush resistance and protection against abrasion, are wound around the cable in a spiral. The result of this wrapping technique is that this additional layer does little to assist the strength members, and certainly could not replace them. As we know, if we pull on a spiral, such as a spring, it gets longer. Applied to a cable, it would mean that pulling forces would cause the cable to lengthen until the fibers break.

The cost of the optic fiber within the outdoor cable is an insignificant part of the whole. It pays then to install considerable spare capacity in the form of extra optic fibers to allow for failures or expansion of the traffic carried in the future. The spare fibers do not degrade by laying around unused but to add an extra fiber after installation would mean laying a whole new cable.

Rodents

One form of mechanical damage which may need to be addressed in some situations is rodent damage. In both fiber and electrical cables they gnaw away until they either break through the optic fiber or they reach the live conductor and the cable takes its revenge.

Rodent attack can be lessened by using a very hard plastic for the outer sheath or, more usually, relying on the layer of galvanized wire strands that was added for general mechanical protection. An alternative approach is to install the cable in a rodent-proof duct.

Installation tension

As the name suggests, this is a measure of how hard we can pull the cable during installation without causing damage. The force is measured in Newtons (N). The quoted values start at about 100 N for lightweight internal cables and extend to 10,000 N or more for ruggedized cables.

So what does all this mean? How big is a Newton?

For those of us who enjoy the unlikely story of Sir Isaac Newton sitting under an apple tree, it may be interesting to know that to pick up an average sized eating apple he would have applied a force of about 1 N. An equally unscientific measure is achieved by taking off your jacket, rolling up your sleeves, taking a firm grip on the cable and pulling hard until you break out in a gentle sweat. We are now looking at a force in the region of 250 N. Slightly nearer the mark is to remember that about 10 N is needed to lift a 1 kg (2.2 lb) weight.

Both eating apples and people come in a wide range of sizes. For this reason, the apparatus for pulling the cables is fitted with adjustable tension and mechanical fuses to prevent damage to the cable.

Weight

Expressed in kilograms per kilometer ($kgkm^{-1}$). There is obviously a wide range of weights depending on the degree of protection built into the cable. Light duty indoor cable designs with the Kevlar strength members result in weights around 20 $kgkm^{-1}$ and the armored external cables are in the order of 500 $kgkm^{-1}$. Hybrid cables which include copper cores can be very much heavier.

Bending radius

To prevent damage to the cable and possible bending losses, the minimum bend radius is always quoted in the specification. One interesting point is that two different figures are stated, a tighter bend is allowed for long term use since the cable is no longer under stress once installation is complete. For a lightweight cable the figures are about 50 mm for long term use and 70 mm during installation, and for external armored cable the figures are around 175 mm and 350 mm respectively.

Temperature range

Three temperatures may be offered in a specification:

- ▷ installation range typically 0°C to +60°C
- ▷ operating range, –30°C to +70°C
- ▷ storage range, –40 to +80°C

The installation range is a little more restricted because the cable will be under stress during this time.

Typical examples

To bring it all together, here are two typical cables. The first is a lightweight internal tight jacketed and the other is an external armored cable suitable for direct burial.

Lightweight internal cable — Figure 8.7

Cable diameter	4.8 mm
Minimum bending radius	
long term	40 mm
installation	60 mm
Installation tension	
long term	250 N
installation	800 N
Weight	19 $kgkm^{-1}$
Temperature range	
installation	0°C to +60°C

static operation	–10°C to +70°C
storage	–20°C to +80°C

External direct burial cable — Figure 8.9

Cable diameter	14.8 mm
Minimum bending radius	
long term	150 mm
installation	225 mm
Installation tension	
long term	600 N
installation	3 000 N
Weight	425 kgkm[1]
Temperature range	
installation	0°C to +60°C
static operation	–10°C to +60°C
storage	–20°C to +70°C

Installation of the cable

Most fiber is installed by traditional pulling techniques developed for copper cable. A pulling tape is inserted into the duct and is attached to the end of the optic fiber cable by a cable grip. This is a length of stainless steel braided wire terminated by an eye. As the tape is pulled, the tension increases and the braiding decreases in diameter gripping the cable somewhat like a Chinese finger trap. The harder we pull, the tighter it gets. To reduce the friction experienced by the cable as it slides along the duct proprietary lubricants can be applied to the outside of the cable.

Blown fiber

The is an alternative technique available for distances up to about 2 km. It is unique in that it involves the installation of a cable empty of any optic fibers and these are added afterwards. This has three real advantages. It defers much of the cost since only the optic fibers actually required at the time need to be installed. It allows the system to be upgraded by the replacement of individual optic fibers. Finally, as a bonus, the installation of the optic fiber is completely stress free.

The first step is to install the empty loose tube cable. The cable shown in Figure 8.10 contains seven tubes each of 6 mm inside diameter in an outer jacket of 28 mm diameter. No commitment to use any particular number or types of fibers need to be made at this stage in the proceedings.

The next stage is to take a prepared bundle of fibers, usually four, contained in a tight jacket of foamed polyethylene. This jacket of foamed polyethylene produces a white coating which is very light in weight and slippery to the touch. It is reminiscent of a slippery version of expanded polystyrene. Its diameter of about 3 mm is a very loose fit inside the tubes that we have installed in the duct.

A small compressor, called a blowing head, is attached to the end of the loose tube and blows air through it. The fiber is then fed in through the same nozzle as the air and it is supported by the airstream. The fiber is blown along the tube and happily round bends in the duct, like a leaf in the wind, at a rate of about two meters a second, a goodish walking pace. No stress is felt by the fiber as it is supported by the movement of the air all along its length.

Other fibers can be blown in at any time to suit the customer. The fibers are equally easy to remove if the fiber needs to be upgraded at a future date.

Only one fiber bundle can be installed in each tube so the system can consist of seven tubes each containing a single bundle of four fibers, twenty eight fibers in all.

The manufacture of optic fiber

This is intended as a brief outline to the process offered more out of interest than any pretense of it being useful.

There are several different methods in use today but the most widespread is called the modified chemical vapor deposition (MCVD) process.

At the risk of oversimplifying a complex process, we can summarize the process in three easy stages.

Stage 1 — Figure 8.11

A hollow silica tube is heated to about 1500 °C and a mixture of oxygen and metal halide gases is passed through it. A chemical reaction occurs within the gas and a glass 'soot' is formed and deposited on the inside of the tube. The tube is rotated while the heater is moved to and fro along the tube and the soot forms a thin layer of silica glass. The rotation and heater movement ensures that the layer is of constant thickness. The first layer that is deposited forms the cladding and by changing the constituents of the incoming gas the refractive index can be modified to produce the core. Graded index fiber is produced by careful continuous control of the constituents.

Stage 2 — Figure 8.12

The temperature is now increased to about 1800 °C and the tube is collapsed to form a solid rod called a preform. The preform is about 25 mm in diameter and a meter in length. This will produce about 25 km of fiber.

Stage 3 — Figure 8.13

The preform is placed at the top of a high building called a pulling tower and its temperature is increased to about 2100 °C. To prevent contamination, the atmosphere is kept dry and clean. The fiber is then pulled as a fine strand from the bottom, the core and cladding flowing towards the pulling point. Laser gauges continually monitor the thickness of the fiber and automatically adjust the pulling rate to maintain the required thickness. After sufficient cooling, the primary buffer is applied and the fiber is drummed.

Figure 8.10
Blown fiber tubing
— no fiber installed

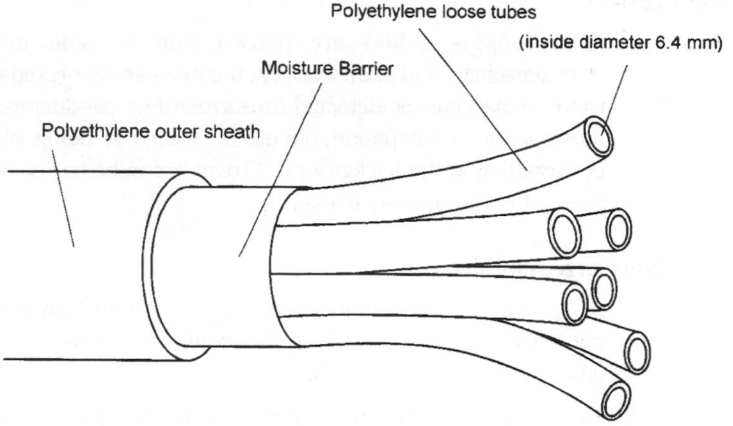

Polyethylene loose tubes
(inside diameter 6.4 mm)

Moisture Barrier

Polyethylene outer sheath

Figure 8.11
Stage 1 — the core
and cladding are
formed inside the
tube

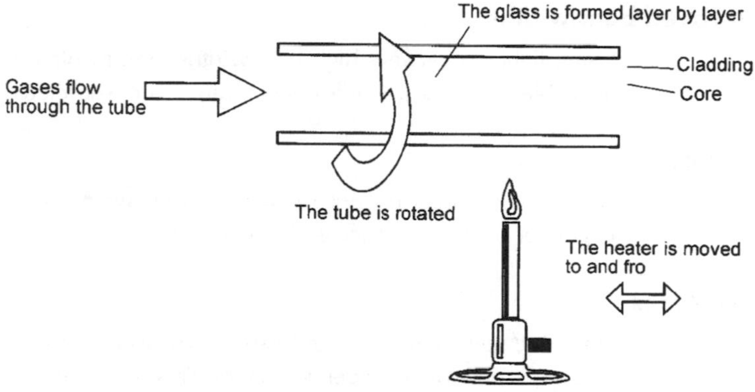

The glass is formed layer by layer

Cladding
Core

Gases flow
through the tube

The tube is rotated

The heater is moved
to and fro

Figure 8.12
Stage 2 — further
heating collapses
the tube

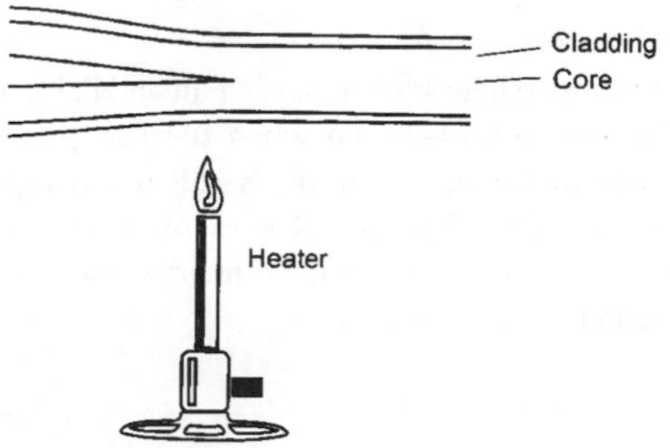

Cladding
Core

Heater

Advantages of optic fibers

The fact that optic fibers do not use copper conductors and even the strength members need not be metallic gives rise to some advantages.

Immunity from electrical interference

Optic fibers can run comfortably through areas of high level electrical noise such as near machinery and discharge lighting.

No crosstalk

When copper cables are placed side by side for a long distance, electromagnetic radiation from each cable can be picked up by the others and so the signals can be detected on surrounding conductors. This effect is called *crosstalk*. In a telephone circuit it results in being able to hear another conversation in the background. Crosstalk can be easily avoided in optic fibers even if they are closely packed.

Glass fibers are insulators

Being an insulator, optic fibers are safe for use in high voltage areas. They will not cause any arcing and can be connected between devices which are at different electrical potentials.

The signals are carried by light and this offers some more advantages.

Improved bandwidths

Using light allows very high bandwidths. Bandwidths of several gigahertz are available on fibers whereas copper cables are restricted to about 500 MHz.

Security

As the optic fibers do not radiate electromagnetic signals, they offer a high degree of security as discussed in Chapter 6.

Low losses

Fibers are now available with losses as low as 0.2 dBkm^{-1} and hence very wide spacing is possible between repeaters. This has significant cost benefits in long distance telecommunication systems, particularly for undersea operations.

Size and weight

The primary coated fiber is extremely small and light, making many applications like endoscopes possible. Even when used as part of a cable with strength members and armoring, the result is still much lighter and smaller that the copper equivalents. This provides many knock-on benefits like reduced transport costs, more cables can be fitted within existing ducts and they are easier to install.

Figure 8.13

Stage 3 — the fiber

is drawn

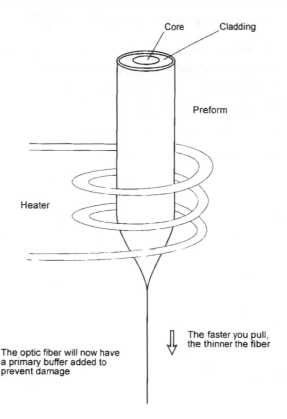

The optic fiber will now have
a primary buffer added to
prevent damage

The faster you pull,
the thinner the fiber

Quiz time 8

In each case, choose the best option.

1   A cable containing both optic fiber and copper conductors is called:

(a) an armored cable

(b) a dehydrated cable

(c) tight jacketed cable

(d) a hybrid cable

2   LSOH means low smoke:

(a) zero halogen

(b) zero heat

(c) optic fibers of hybrid design

(d) or heat

3   Spiral galvanized wires are sometimes added to a cable to:

(a) conduct electricity

(b) allow the cable to stretch more to relieve stress

(c) improve crush resistance

(d) allow the fibers to be upgraded as necessary

4   Blown fiber allows:

(a) easy removal of the fibers to clean the ducts

(b) the diameter of the fiber to be increased by filling it with compressed air

(c) easy removal of any water in the fiber

(d) easy replacement of any damaged fiber

5   Within buildings:

(a) both tight buffered and loose jacketed cables are used

(b) only loose jacketed cables are used

(c) only armored cables can be used

(d) only tight buffered cables are used

9

Connecting optic fibers — the problems

There are only three real problems involved in connecting optic fibers:

- ▷ the fibers must be of compatible types
- ▷ the ends of the fiber must be brought together in close proximity
- ▷ they must be accurately aligned.

Compatibility problems

A curious feature of these compatibility problems is that they result in the degree of loss being dependent on the direction of travel of the light along the fiber. Hitherto, we have always considered light direction to be irrelevant.

Core diameters

Multimode fibers come in a wide variety of core sizes between 7 µm and 2 mm, of which the most usual are 50 µm, 62.5 µm, 85 µm, 100µm, 200 µm. Similarly, the all-plastic fibers range from 0.25 mm to 3 mm of which 1 mm is the most common.

When we purchase components, such as a laser, they often come attached to a length of optic fiber (called a *pigtail*) which we cannot disconnect. If this fiber has characteristics different to those of our main system, then we must be aware of a possible power loss at the point of connection.

Single mode fibers are restricted to a size very close to 8 µm, so core size problems are not common.

If we connect a multimode fiber with a large core to one with a smaller core, as shown in Figure 9.1, only some of the light emitted by the larger core will enter the smaller core due to the reduced area of overlap and a power loss will occur. If, however, the light traveled from the smaller core to the larger, all the active core is overlapped and no losses will occur.

Figure 9.1

Losses due to

unequal core sizes

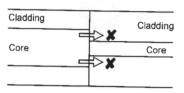

Some light cannot enter the core

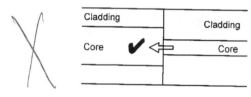

Small core to large core - no losses

		Launch fiber core size (µm)		
		9	50	62.5
Receive fiber core size (µm)	9	0	14.8 dB	16.8 dB
	50	0	0	1.9 dB
	62.5	0	0	0

Losses can be high

The size of the loss can be calculated using the formula :

$$\text{Loss} = -10\log_{10}\left(\frac{\text{core diameter}_{\text{receive}}}{\text{core diameter}_{\text{launch}}}\right)^2 \text{ dB}$$

Remember that this formula only applies when the diameter of the launch fiber is greater than that of the receiving fiber otherwise there are no losses.

Numerical apertures

A very similar effect occurs with changes in the numerical apertures. If the receiving fiber has a numerical aperture which is equal to, or greater than, the launch fiber, no losses will occur.

The reasoning behind this is that the numerical aperture determines the cone of acceptance.

Let us assume a typical fiber with a numerical aperture of 0.2, resulting in a cone of acceptance of 11.5° is connected to a fiber with NA = 0.25, with a cone of acceptance of 14.5° (Figure 9.2). Using this direction of transmission, all of the ray angles would be accepted by the other fiber and no losses would occur. If, however, we transmitted light in the other direction, the light rays with angles between 11.5° and 14.5° are outside the cone of acceptance of the receiving fiber would not be accepted and would result in a loss.

Figure 9.2

Losses due to

changes in

numerical aperture

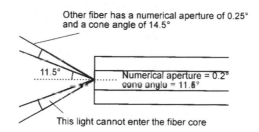

Other fiber has a numerical aperture of 0.25°
and a cone angle of 14.5°

11.5°

Numerical aperture = 0.2°
cone angle = 11.5°

This light cannot enter the fiber core

Launch fiber NA

		0.1	0.2	0.275
Receive fiber NA	0.1	0	6 dB	0.36 dB
	0.2	0	0	2.8 dB
	0.275	0	0	0

Example losses

The formula looks very similar to that used for the problem of the core diameters:

$$\text{Loss} = -10\log_{10}\left(\frac{NA_{receive}}{NA_{launch}}\right)^2 \text{ dB}$$

Remember that the formula only applies when the numerical aperture of the launch fiber is greater than that of the receiving fiber. Otherwise no losses occur.

What if both the core sizes and the numerical apertures differ?

Good news!

We can find the total loss by simply adding the two separate losses.

Example

What loss would result from connecting a fiber with a core size of 62.5 µm and NA of 0.275 to a fiber of core size 50 µm and NA of 0.2?

From the formulae, or reading off the tables, we obtain a loss due to core size of 1.9 dB and a NA loss of 2.8 dB.

The total loss will be: 1.9 + 2.8 = 4.7 dB

Gap loss — Figure 9.3

As the ends of the fiber are separated, the light from the core spreads out at an angle equal to the acceptance angle as we saw in Chapter 4. Less light strikes the core area of the receiving fiber and a loss occurs.

The degree of loss is not severe with a value of less than 0.5 dB when the ends of the fiber are separated by a distance equal to a core diameter. The loss is reduced even further by the use of index matching gel which is added in the joint to make the fiber core appear continuous. Index matching gel has a refractive index similar to the core of an optic fiber and is used to fill the gap between the fibers to make the light path appear to be continuous. It is normally added to reduce fresnel reflections but it helps with gap loss at the same time.

Gap loss increases linearly with the size of the gap.

Alignment problems — Figure 9.4

Lateral misalignment

This has a certain similarity to the loss due to differences in core size. As the fibers are moved the area of overlap between the two cores is reduced and hence less light transfer occurs. This alignment is quite critical, much more so than gap loss, and a misalignment of one quarter of a core diameter will cause a loss of 1.5 dB. Thereafter the loss increases rapidly in a non-linear manner.

Core alignment (eccentricity loss)

This happens when the core is not positioned exactly in the center of the fiber. When optic fibers are connected, they are normally aligned by reference to the outside of the cladding. If the core is not placed centrally within the fiber, the result will be misalignment of the core giving results similar to those caused by lateral misalignment. Using modern manufacturing techniques, serious examples of this fault are rare.

Angular misalignment

As the angular displacement increases, the light from one core progressively misses the other. With an air gap, angles of misalignment up to 3 or 4 degrees cause losses of less than 1dB. At increased angles, the losses increase at faster rates in a nonlinear fashion. The use of index matching gel actually makes the situation worse as it prevents the spreading effect of the cone of acceptance which would normally occur in an air gap.

Connecting optic fibers — the preparation

Whatever method we choose to connect fibers we need to prepare the fibers by stripping off the primary buffer to expose the cladding. The cladding is used to position the fibers to prevent misalignment losses and, ignoring eccentricity

Figure 9.3

Gap loss

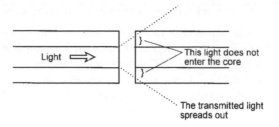

Light ⟹

This light does not
enter the core

The transmitted light
spreads out

Figure 9.4

Angular

misalignment

The fibers are misaligned

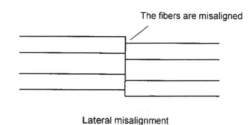

Lateral misalignment

The cladding is aligned

The core is not central

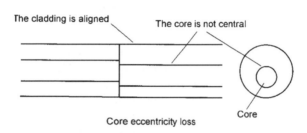

Core eccentricity loss

Core

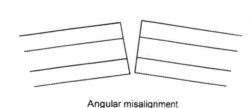

Angular misalignment

loss, ensures that the cores are aligned. The primary buffer is not sufficiently accurate or stable for this purpose.

A safety point

Once the primary buffer is removed, the core and cladding are, of course, very thin. This means that a gentle pressure applied to the fiber will translate into extremely high concentrations of force over the end area of the fiber, easily puncturing the skin, often without being felt if it misses a nerve ending. Worse than that, it can then break off inside your hand. We have yet to discover the long-term medical effects but it is certainly not going to prove beneficial.

Stripping

The primary buffer is first exposed by removing the outer jackets. This may mean a few seconds with a simple wire stripper tool such as is used for electrical wiring or an hour fighting your way through layers of armoring.

There are two approaches to stripping off the primary buffer — chemical or mechanical. There is a variety of plastics used for this buffer and the suppliers will advise on the best method. Whatever the method used, the aim is to remove about 50 mm of the buffer without scratching or nicking the surface of the cladding as this would cause the fiber to break.

Chemical

The advantage of the chemical method is that there is no danger of any mechanical damage to the cladding.

A small bottle of stripper is supplied. The liquid or gel is brushed onto the surface of the buffer and left for two or three minutes. It softens the plastic which can then be wiped off with a tissue. The stripper smells remarkably like paint stripper and gives off equally unpleasant fumes. It should be treated with respect, ensuring that it does not come into contact with eyes, skin or naked lights. Use finger stalls to avoid contact with the skin and safety glasses. Remember, we cannot see with glass eyes.

Mechanical

There is a range of hand tools that can provide the same service without the use of chemicals, an example is shown in Figure 9.5. Their operation is basically the same as copper wire strippers except that they are built with much greater accuracy. The principle is to cut through the buffer with a blade stopping just short of touching the surface of the cladding. The fiber is then pulled through the jaws and the blade tears off the primary buffer. Because of the precision of construction, the tools are made for specific buffer sizes.

Figure 9.5

Remove the
primary buffer

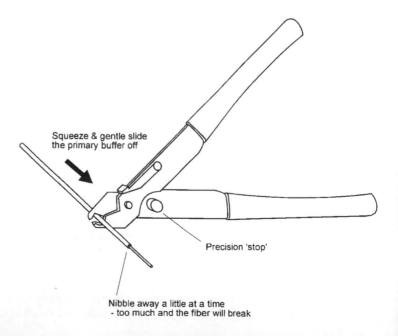

Squeeze & gentle slide
the primary buffer off

Precision 'stop'

Nibble away a little at a time
- too much and the fiber will break

Cleaning

Every trace of the buffer or other contamination must be carefully removed. A single speck of buffer, dust or grease will cause the misalignment of fibers and the consequent loss of light.

The most usual method of cleaning is to moisten some lint free cloth or tissue with isopropyl alcohol (IPA), fold it over and pull the fiber through the tissue. When the fiber is clean it gives off a squeal so is, quite literally, squeaky clean. It is good to avoid too much contact with IPA and the use of fingerstalls are a good idea. Acetone is sometimes offered as an alternative cleaning fluid but the health hazards of skin contact and breathing the fumes are considered worse (even worse) than isopropyl alcohol. Mixing cleaning fluids can be dangerous so choose one and stay with it.

If the cable is gel-filled, it is essential that the gel is not allowed to contaminate the tools or the cladding. Turpentine or white spirit can be used as a cleaning agent at this stage.

Once the fiber is completely clean, ensure that it is not recontaminated. This is so easily done, almost without a thought. Some people, having cleaned the fiber, feel an overwhelming urge to pull it through between their finger and thumb just to make sure, others are careful to avoid this and carefully lay the fiber down on the dusty work surface!

Cleaving

To connect two fibers, as well as being stripped and totally clean, the end of the fiber must be cut cleanly at right angles. This process is called cleaving. We are looking for the error in this angle to be no more than 1°. Any greater error will give rise to angular losses. We want the fiber to break very cleanly leaving the end face with no flaws.

In Chapter 4, when we were discussing the need for a primary buffer, we saw how cracks could be propagated through the fiber by introducing a small nick in the surface and then introducing some stress in the fiber. We use exactly this process for cleaving the fiber.

There are many different cleaving tools on the market ranging in size, (claimed) accuracy, and price. The price difference between the top and bottom of the range is in the order of thirtyfold. Generally, the small inexpensive ones tend to be a little more difficult to use and require practice to obtain the 'knack'. The more expensive ones are easier to use giving quality cleaves every time with the minimum of training. It is a good idea to try several types perhaps at an exhibition or when attending a training course — there is always a range of opinion as to which is best. The 'best' one is probably the one you feel most happy using.

The perfectly clean fiber is supported in position, often by a vee-groove as can be seen in Figure 9.6. If a vee shaped groove is accurately cut in a stable material the fiber will fall to the bottom and will align itself with the groove. Any contamination must be avoided at all costs — both the vee-groove and the fiber must be scrupulously clean otherwise it will prove disastrous.

A blade is brought up to the surface of the cladding and makes a minute mark

Figure 9.6

The vee-groove
method of
positioning optic
fiber

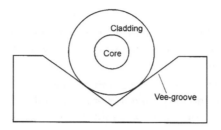

The vee-groove positions the fiber with precision

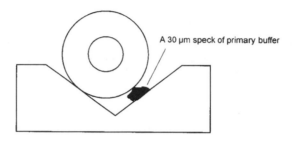

The slightest contamination can ruin the effect

or witness on the surface (Figure 9.7). The blade must be at 90° to the fiber and the mark should be no more than 5 µm deep.

In the more sophisticated cleavers there is considerable research and engineering precision employed in achieving this and it is fully automatic. At the cheap end of the market a hand tool is provided consisting of an artificial sapphire mounted in a pen type holder. The operator must support the fiber against one finger as the 'pen' is stroked gently across the fiber. It takes much practice to produce the correct mark on the cladding. The results are, quite literally, left in the hands of the operator.

Once the witness, or mark, is made, the fiber is stressed by being pulled and the crack propagates across the fiber and, hopefully, it is cleaved. The waste fiber must be disposed of in sealed containers.

Inspection of the cleave

The end condition is checked using a microscope or a magnifying glass. A watchmakers' eyeglass is useful for this. We are interested in the end angle, and checking for any tangs, chips or cracks (Figure 9.8). If any of these are present it will prevent efficient light transfer and there is no alternative but to start the cleave again.

Plastic fiber

The outer jacket is removed with a pair of wire strippers, then the end of the fiber is ready for cleaving. The cleaving process is really simple — the fiber is simply cut with a sharp knife. Things are not so critical with plastic nor the results so good.

Figure 9.7
The cleaving
process

Figure 9.8
Unacceptable
cleaves

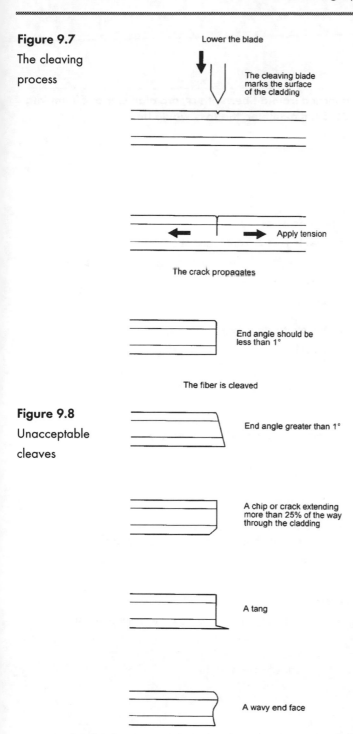

Lower the blade

The cleaving blade
marks the surface
of the cladding

Apply tension

The crack propagates

End angle should be
less than 1°

The fiber is cleaved

End angle greater than 1°

A chip or crack extending
more than 25% of the way
through the cladding

A tang

A wavy end face

Quiz time 9

In each case, choose the best option.

1   If light is launched from a fiber with a core diameter of 50 μm into a fiber of core 62.5 μm diameter, the loss would be:

(a) +1.9 dB

(b) zero

(c) −1.9 dB

(d) 16.8 dB

2   Cleaving is the process of:

(a) removing the cladding before connecting fibers together

(b) cutting the end of the fiber in preparation for connecting two fibers

(c) cleaning the surface of optic fibers

(d) inspecting fibers for flaws

3   A tang is most likely to give rise to:

(a) increased values of NA

(b) eccentricity

(c) excessive amplification

(d) gap loss

4   The core of a typical monomode fiber would have a diameter of:

(a) 50 μm

(b) 50 nm

(c) 8 μm

(d) 1 mm

5   Stripping off the primary buffer can be achieved by:

(a) chemical or mechanical means

(b) cleaving the buffer off

(c) isopropyl alcohol

(d) brushing on index matching gel then wiping off with lint free cloth

10

Fusion splicing

Fusion splicing is the most permanent and lowest loss method of connecting optic fibers. In essence, the two fibers are simply aligned then joined by electric-arc welding. The resulting connection has a loss of less than 0.1 dB (less than 2% power loss). Most fusion splicers can handle both single mode and multimode fibers in a variety of sizes. There are also splicers that can automatically splice multicore and ribbon cable up to 12 fibers at a time.

Preparation of the fiber

The fibers must first be stripped, cleaned and cleaved as we have seen in Chapter 9. To allow spare fiber for easy access and to allow for several attempts, a length of at least five meters of jacket should be removed. The primary buffer is only stripped to about 25 mm. The exact length is determined by the fusion splicer in use.

The quality of the cleave is of paramount importance. However much money we spend on buying the most sophisticated splicing apparatus, it will all be wasted if we cannot cleave the fiber accurately. Both cleavers and splicers come in a range of prices with splicers being the more expensive by a factor of at least ten and sometimes a hundred. It is never a good idea to save money by buying an inadequate cleaver — it is always better to buy the cleaver you have confidence in then, if necessary, recover the money by buying a slightly cheaper version of the splicer. Most splicers nowadays measure the accuracy of the cleave and if found wanting, the fiber is rejected until you have redone it to a satisfactory standard. Most splicers consider an end angle of better than about 3° as satisfactory.

Protecting the fiber

Splice protector

In the preparation phase, we have stripped the fiber of all its mechanical and waterproof protection. Once the fiber has been spliced, some protection must be restored since the splicing process will have reduced the fiber strength to less than 30% of its former value.

This is achieved by a device called a splice protector. It consists of a short length (about 60 mm) of heatshrink sleeving enclosing some hot-melt glue and a stainless steel wire rod as seen in Figure 10.1.

Figure 10.1

Don't forget the

splice protector

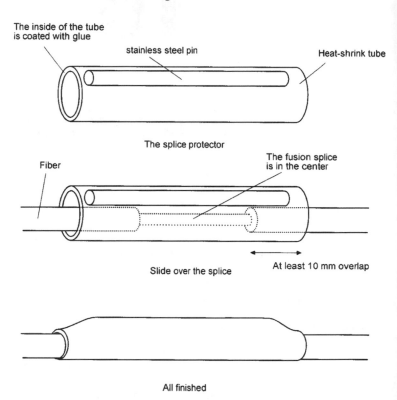

The inside of the tube is coated with glue

stainless steel pin

Heat-shrink tube

The splice protector

The fusion splice is in the center

Fiber

Slide over the splice

At least 10 mm overlap

All finished

Prior to joining the fiber, the splice protector is slid onto the fiber. After the splice is completed, the protector is centered over the splice and heated, usually in a purpose-built oven although a hot-air gun can be used. The oven is a simple tray with a lid, a heater and a timer which are normally built-in features of the splicers. The hot-melt glue keeps the protector in position whilst the stainless steel rod provides proof against any bending that may occur. The outer sleeve offers general mechanical and water protection to replace the buffer that has been removed. To ensure that the fiber is fully protected along its length, at least 10 mm of the protector must overlap the primary buffer at each end of the splice.

Enclosures (termination enclosures)

After the splice is completed, we are left with a length of fiber deprived of its outer jacket. The fiber must be protected from mechanical damage, and from water. This is achieved by an enclosure (Figure 10.2). The design, and cost, of the enclosures depend on the environment in which the fiber is going to live.

Obviously something to protect fiber under water has to be superior to a plastic box in an air-conditioned office.

Figure 10.2

A very simple splice enclosure

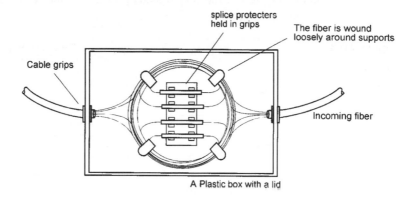

splice protecters held in grips

The fiber is wound loosely around supports

Cable grips

Incoming fiber

A Plastic box with a lid

The significant feature is a means of ensuring that the fiber is well supported within the container in such a way that bending loss is avoided. This is done by having something to wind the fiber around, like a reel, referred to as a cassette, or at least a few clips to support the fiber and the splices.

They are readily available in different sizes to hold everything from 4 to 240 fibers. Each fiber must be identified, otherwise a simple job could become a real nightmare. This is achieved by attaching labels to the fibers or splice protectors and by using colored splice protectors.

There are some other factors to consider which may not immediately spring to mind such as:

> security of the data. With the outer jacket removed a simple live fiber detector can be clipped onto the fiber and all the data being passed can be copied.

> access to an enclosure is the easy way to sabotage a communication system.

> there is also a problem with light, again with no jacket and as bends are inevitable, there is a risk of light entering the system so the container should be light proof.

> unpleasant environments. Salt spray, acids, high temperatures, crushing and all sorts of other nasties.

> access for repairs or for testing purposes.

We stripped off five meters of outer jacket to enable the fiber to be lifted out of the enclosure with enough spare fiber to be easily connected to test equipment or a fusion splicer.

Holding and moving the fibers in the splicer

The fibers are held in vee-grooves cut into steel or ceramic blocks (Figure10.3). As usual, cleanliness is all-important. The fiber is cleaned and cleaved then the vee-groove is cleaned by a lint free cloth, tissue or a 'cotton bud' moistened with isopropyl alcohol. Do not use compressed air cleaners as any contamination will turn it into a grit blaster and damage the critical dimensions of the vee-groove. The fiber is gently pressed into the vee-groove by a magnetic or gravity clamp.

Figure 10.3

The fiber is held in

position

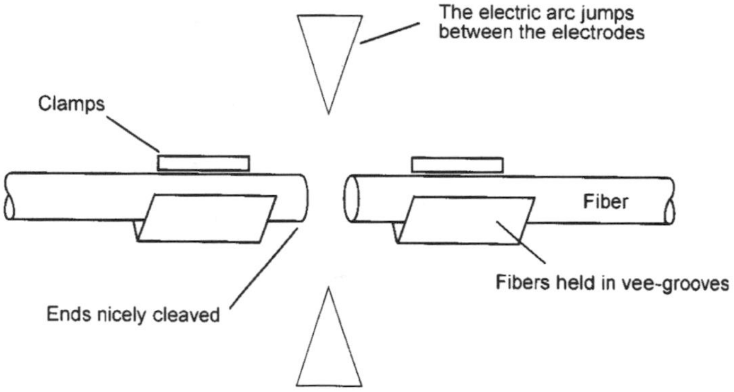

The electric arc jumps between the electrodes

Clamps

Fiber

Fibers held in vee-grooves

Ends nicely cleaved

Once the fibers are safely clamped into their vee-grooves, they are moved, vee-grooves and all, until the fibers are aligned with each other and positioned directly under the electrodes from which the electric arc will be produced. We are aiming to achieve positioning with an accuracy of better than 1 μm.

In the least accurate splicers, suitable only for multimode fibers, this can be achieved by simple microgears operated manually. More precision is required for single mode fiber since the core is so much smaller. A 1 μm error in positioning an 8 μm core causes a lateral misalignment of 12.5% whereas the same error on the larger 62.5 μm core in a multimode fiber would represent a lateral misalignment of less than 2%.

The extra precision is provided by using stepper motors. A stepper motor is an electric motor that behaves in a different manner to a 'normal' electric motor. We usually picture electric motors spinning round as power is applied. Stepper motors, instead, turn a set number of degrees and then lock in that position. The amount it turns can be precisely controlled by digital input signals and, in conjunction with a gear train, is able to provide extremely accurate alignment of the fibers.

Observing the alignment

All fusion splicers are fitted with some means to observe the fiber positioning and the condition of the electrodes. This is achieved by either a microscope or by a CCD camera (CCD = charge coupled device — a semiconductor light sensor) and a liquid crystal display (LCD). The trend is towards CCD cameras since they are more pleasant to use and have the safety advantage of keeping our eyes separated from the infrared light which can, of course, cause irreparable damage to the eyes if we accidentally observe an active fiber through the microscope.

The optic system always allows viewing from two angles as fibers can otherwise hide one behind the another and appear to be aligned. One way of achieving this is shown in Figure 10.4.

Figure 10.4

Observing the fiber positions

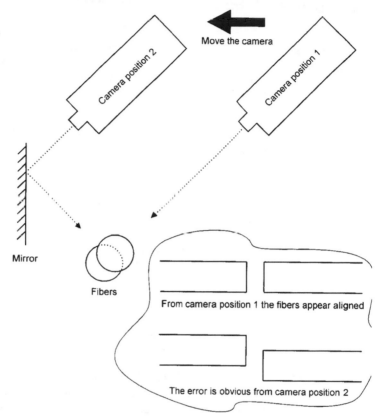

Automatic positioning

There are two methods.

PAS — the profile alignment system — Figure 10.5

This is the standard method of aligning the fibers in modern fusion splicers. The idea is very simple. A light is shone through the fiber and is detected by a CCD camera. The change of light intensity at the edge of the cladding and at the core due to the changes in refractive indices allows the system to detect their positions. Several readings are taken from each fiber and averaged out to reduce any slight errors.

Once the positions are detected, small stepper motors are activated to bring the two fibers into alignment. The viewing angle is switched through 90° to allow the system to check in both planes and further small adjustments are made until the splicer is quite satisfied.

The whole operation is usually automatic but we can follow the process on the liquid crystal display. As the system is able to detect the core position as well as the cladding, any eccentricity error in the core can be compensated for.

Figure 10.5

PAS — detecting the

position of the fiber

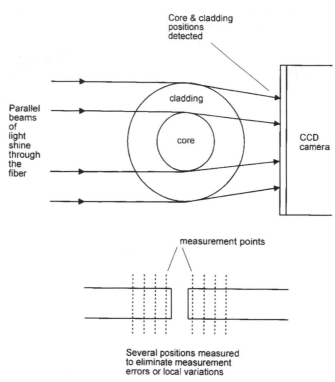

Core & cladding positions detected

Parallel beams of light shine through the fiber

cladding

core

CCD camera

measurement points

Several positions measured to eliminate measurement errors or local variations

LID — light injection and detection system — Figure 10.6

This system makes use of bend loss, the light leakage that occurs when moderately tight bends are introduced into a fiber. Remember that the light can go into the fiber at a bend as well as being able to escape from it.

A bend is introduced at the input side of the splicer and light is injected into the fiber through the primary buffer. The light travels down the fiber and jumps the gap into the other fiber. A similar bend in the other length of fiber allows light to escape.

A stepper motor is used to move one of the fibers horizontally and the output light is monitored to detect the point of maximum light transfer. This means that the cores are aligned, at least in one plane. The fiber is then moved in the vertical plane until, once again, the point of maximum light transfer is discovered. The whole process is repeated once or twice making finer and finer adjustments until it homes in on the point of best light transfer. Once this has been achieved, the fibers are spliced.

There are one or two slightly worrying aspects with this design. The first is the severity of the bends introduced. The radius of these bends is generally tighter than the fiber specification allows. This means that if the fiber fails, immediately or at a later date, the fiber manufacturer will not be interested since you exceeded the fiber limits. It does not usually break, of course, but there is still a slight feeling of unease. There is another thought, too. If the primary buffer

Figure 10.6

A simplified LID system

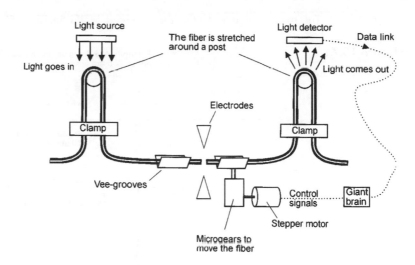

happens to be opaque, the light cannot penetrate. LID systems are sometimes offered as a bolt-on goody to the standard PAS splicer.

Fusion splicing of the fibers

The arc that occurs between the two electrodes is about 7000 volts with an adjustable current up to 25 mA — a very unpleasant and dangerous combination.

Safety precautions are built into splicers to ensure that we cannot accidentally come into contact with the arc. There are two approaches to this problem, the first is to be sure we know where our fingers are before the arc starts. This is done by involving two buttons both of which must be held down during the operation and are spaced so that both hands must be used. The alternative is to enclose the dangerous area with a cover that must be closed before the arc can be energized.

Prefuse

The first stage of the fusing process is to align the fibers with an end gap equal to one fiber diameter and apply a short, relatively low power arc. This is called a prefuse. Its purpose is to clean and dry the end surfaces of the cleaves so that nothing untoward gets trapped inside the splice. It can remove very slight tangs from the cleave but don't expect miracles, it won't repair a poorly prepared fiber.

Main fuse

The fibers are then brought together and some additional end pressure is applied. The additional pressure allows the fibers to move towards each other slightly as they melt. How far they move, called the overfeed or stuffing distance is critical. Too much or too little and the splice will not be satisfactory.

The reason for introducing the overfeed is illustrated in Figure 10.7. In the

Figure 10.7

The need for some

overfeed

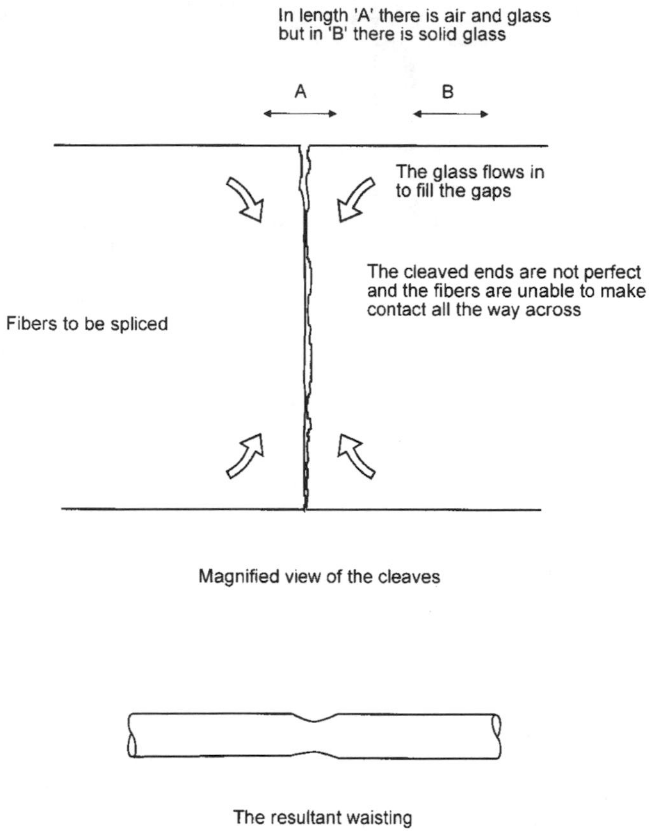

In length 'A' there is air and glass
but in 'B' there is solid glass

A   B

The glass flows in
to fill the gaps

The cleaved ends are not perfect
and the fibers are unable to make
contact all the way across

Fibers to be spliced

Magnified view of the cleaves

The resultant waisting

figure, the cleaved end of the fiber is shown greatly magnified and, even with a good cleaver the surface is never completely smooth. When two fibers are brought together there are some small air gaps present which means that within the area marked A there is actually less glass than in a similar length, shown as B. When the electric arc melts the end of the fiber, the glass tends to collapse inwards, filling the air gaps. This produces some waisting as shown on the page of disasters (see Figure 10.9).

The main fusing arc is more powerful and lasts for a longer period of time, between 10 and 20 seconds.

If the splicer is fitted with a microscope, it is important to check in the instruction book to see if the fiber can be viewed during splicing. The arc emits ultraviolet light and as the microscope concentrates the light there is a danger of suffering from arc-eye, a temporary or permanent damage to the eyes. This is the result of electric arc welding without eye protection. Most viewing microscopes have a UV filter (and an infrared filter) to prevent eye damage, but check the manual first. This is another advantage of choosing one with a CCD camera and monitor.

Once fusing is completed, have a good look at the splice. If it is difficult to see where the splice is, then it's probably a good one (Figure 10.8). We are looking

Figure 10.8

This is what we

hope to see

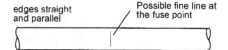

edges straight
and parallel

Possible fine line at
the fuse point

for the outer edges of the cladding to be parallel, just like a new continuous length of fiber. Sometimes a small white line appears across the core but this is not important and can be ignored. See Figure 10.9 for all the disasters.

Figure 10.9

Splicing disasters!

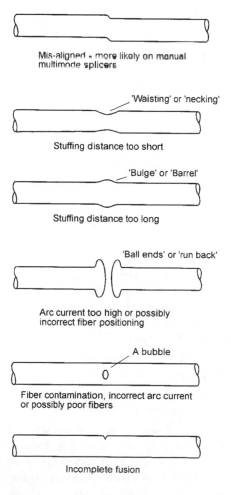

Mis-aligned - more likely on manual
multimode splicers

'Waisting' or 'necking'

Stuffing distance too short

'Bulge' or 'Barrel'

Stuffing distance too long

'Ball ends' or 'run back'

Arc current too high or possibly
incorrect fiber positioning

A bubble

Fiber contamination, incorrect arc current
or possibly poor fibers

Incomplete fusion

Setting up the splicer

Different types of fibers require particular values of arc current and lengths of time since this determines the temperature to which the fiber rises. There are several automatic programs installed as well as user-defined settings that we can employ in the light of experience.

The equipment is powered by nicad batteries giving up to 300 splices per charge, somewhat less (50 or so) if the oven is used to shrink the splice protector.

Automatic splice testing

When the splice is completed, the fiber positions are rechecked and an estimated splice loss is calculated and displayed. Although the estimate is reasonably accurate, it is only an estimate and not a measurement. To be really certain, the loss must be measured and this must be done after the splice protector has been fitted just in case the splice is damaged during this operation. Testing is described in Chapter 15.

Some splicers also provide a tensile test — just a gentle pull of 200 g or so. It won't damage the fiber but will test the mechanical integrity of the fiber.

Practical summary — how to do it

1   Consult the instruction books of the splicer and the splice protector to find the recommended stripping lengths of the primary buffer.

2   Strip off the outer jackets and the required length of the primary buffer.

3   Slip on a splice protector.

4   Clean the fiber.

5   Cleave it.

6   If you are not familiar with the splicer, read the handbook. They all differ in the requirements for loading and positioning the fiber.

7   In a typical case, the fibers are inserted into the vee-grooves and moved until the ends of the fibers meet guide lines visible through the lens or camera. Clamp them in position. Have a look at the standard of the cleaves and, if necessary, take the fiber out and try again.

8   Set the splicing program to match the fiber in use. The handbook will provide guidance.

9   Press the start button and leave it to it. The program will run through its positioning and splicing procedure, then stop.

10 Carefully lift it out of the vee-grooves and slide the splice protector along the fiber until it is centered over the splice. Make sure you have at least 10 mm of primary buffer inside the splice protector and place it gently in the oven. On some splicers, even this is done for you.

11 Switch on the oven and in a minute or two it will switch off.

And that's it.

Quiz time 10

In each case, choose the best option.

1   Too much overfeed whilst splicing will cause:

(a) waisting

(b) a bubble being formed due to air being trapped inside the fiber

(c) a bulge in the fiber

(d) runback

2   The main fusing arc is likely to have a duration of:

(a) 15 s

(b) 62.5 ns

(c) 10 20 minutes

(d) less than a second

3   PAS stands for:

(a) position adjustment system

(b) profile alignment standard

(c) plane alignment system

(d) profile alignment system

4   For viewing the splicing process, a CCD camera is preferable to a microscope because:

(a) higher magnification can be achieved

(b) better resolution is possible

(c) there is no risk of eye damage

(d) a CCD camera can operate at very low light levels

5   A splice protector:

(a) provides more flexibility at the mechanical joint

(b) is not needed inside an enclosure

(c) protects against unauthorized copying of data

(d) provides mechanical protection to the fusion splice

11

Mechanical splices

The mechanical splice performs a similar function to the fusion splice except that the fibers are held together by mechanical means rather than by a welding technique. Physically, they often look very similar to splice protectors.

Advantages and disadvantages

There are several advantages. They do not require any power supplies. Indeed, many designs require no tools at all beyond a stripper and cleaver, so the mechanical splice can be used in situations that may be considered hostile to many fusion splicers. Mechanical splices are often re-usable and can be fitted in less than a couple of minutes, which makes them ideal for temporary connections.

The disadvantage is that they cause a loss, called the *insertion loss*, of about 0.1–0.3 dB per connection which is significantly higher than a good fusion splice. This would suggest the use of a fusion splice as the first choice in situations where losses are critical.

Cost

On cost grounds alone, the choice between a fusion splicer and mechanical splices depends on the number of splices to be undertaken. If we already have a fusion splicer, the cost of each fusion splice is negligible but for the cost of a reasonably good fusion splicer we could purchase a thousand or more mechanical splices. It is also possible to hire fusion splicers and other fiber optic equipment.

How they work

In essence, it is very easy. The fiber must be stripped, cleaned and cleaved. They must then be aligned and then held in position either by epoxy resin or by mechanical clips.

There are only three basic designs.

Vee-groove

This was the obvious choice since it worked so well in positioning the fibers in the fusion splicer. See Figure 11.1.

Figure 11.1
The principle of most mechanical splices

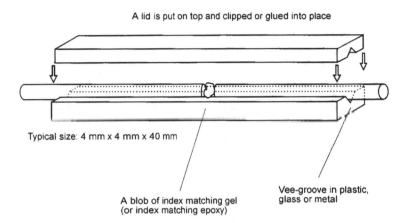

A lid is put on top and clipped or glued into place

Typical size: 4 mm x 4 mm x 40 mm

A blob of index matching gel (or index matching epoxy)

Vee-groove in plastic, glass or metal

Most mechanical splices are designed around the vee-groove. They consist of a base plate into which the vee-groove has been cut, ground or molded.

The prepared fibers are placed in the groove and their ends are brought into contact. Some index matching gel is used to bridge the gap between the two ends to prevent gap loss and to reduce Fresnel reflection. A gripping mechanism then holds the fibers in position and provides mechanical protection for the fiber. As an alternative to the index matching gel, an index matching epoxy can be used. This performs the same index matching task as the gel but also holds the fiber in position. It is usually cured by UV light.

Bent tube — Figure 11.2

If a length of fiber is pushed into a tube which is curved, the springiness of the fiber will force it to follow the outside of the curve. Now, if the tube is of square cross-section, the fiber will follow the far corner. This is very similar to a vee groove since the fiber is now positioned by a vee-shaped wall of the tube. This is called a *bent tube* design. A small spot of index matching gel is added before the fibers are inserted. In some designs, a bent tube with a circular cross section is used but the principle is just the same.

Precision tube

This type is very simple. A hole, very slightly larger than the fiber diameter is formed through a piece of ceramic or other material. When a piece of bare fiber is inserted from each end, the two fibers are inevitably aligned when they meet. The insertion losses are higher than the other types due to tolerances in the hole

Figure 11.2

The bent tube

design

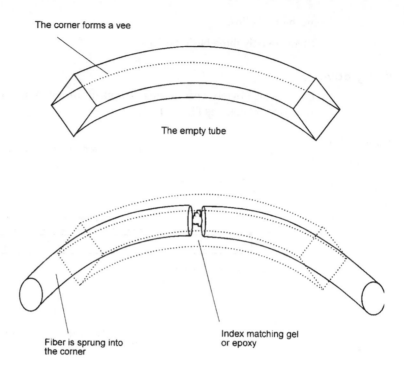

The corner forms a vee

The empty tube

Fiber is sprung into
the corner

Index matching gel
or epoxy

diameter.

An interesting variation on this idea is to use an hole through an elastomeric material as shown in Figure 11.3. An elastomeric material is a group of soft plastics that spring back into shape when deformed.

Figure 11.3

An elastomeric

splice

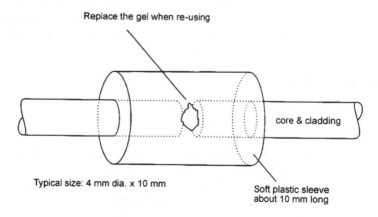

Replace the gel when re-using

core & cladding

Typical size: 4 mm dia. x 10 mm

Soft plastic sleeve
about 10 mm long

A quick and easy splice can be made from this material by simply having a slightly undersized hole through the center. When the fibers are pushed in from each end, they are aligned by the springiness of the elastomer, resulting in an insertion loss of only 0.2 dB.

In the center of the splice there is the usual bead of index matching gel. This form of splice is for temporary connection in the field or the laboratory and is completed in seconds. A syringe can be used to squirt in some new index matching gel and, in this way, it can be re-used up to 50 times. It requires no tools at all once the fiber is stripped and cleaved.

Specifications

The specifications will give information about several things.

Cladding and buffer diameter

The cladding diameter. It will usually be 125 µm for most fibers, a little more tolerance is accepted with elastomeric splices due to the elasticity of the material. The buffer diameter is likely to be either 250 µm or 900 µm depending on whether it is the grip the primary or secondary buffer.

Insertion loss

This is the loss caused by the device when it is installed in the system. Typical values are around 0.2 dB.

Return loss

This is the proportion of the incoming light that is reflected back along the fiber. Usually between –40 dB and –60 dB The low loss is the result of adding the index matching gel to reduce Fresnel reflection.

Fiber retention

How much tension can be applied to the completed splice before the splice fails? The failure mode may be obvious and catastrophic as in the fiber actually becoming disconnected or the ends of the fiber being pulled apart very slightly causing enormous gap loss but no visible damage. Typical values are around 4 N but some are much more rugged with figures up to 180 N when the correct external protection is applied.

A practical guide to fitting a typical mechanical splice

All mechanical splices come with an instruction sheet. It is essential that the time is taken to read it as each splice has a slightly different fixing method. As mentioned earlier, some are permanent fixings using epoxy resin and some are designed to be re-usable. It is not a good idea to pump in epoxy resin first, then settle down to read the instructions!

1 Strip, clean and cleave the fiber leaving about 12 mm of primary buffer removed.

2 If re-using the splice (if this is possible) clean with isopropyl alcohol and a piece of lint free cloth and use a syringe to inject a small bead of index matching gel into the center of the splice.

3 Release the small clip at one end of the splice and insert the fiber until it comes to a stop. Operate the clip to lock the fiber in position.

4   Release the clip at the other end and insert fiber until it comes up against the fiber already loaded. Operate the clip to lock that fiber also.

5   Test the operation and if not satisfactory, the clips can be released to allow the fiber positions to be optimized.

Quiz time 11

In each case, choose the best option.

1   The three basic designs used for mechanical splices are:

(a) vee-groove, bent tube and precision hole

(b) vee-groove, PAS and LID

(c) fusion splice, mechanical splice and enclosures

(d) UV curing epoxy, index matching gel and isopropyl alcohol

2   A typical value for the insertion loss for a mechanical splice is:

(a) –50 dB

(b) 0.2 dB

(c) 12 mm

(d) 3 dB

3   Mechanical splices have the advantage that they :

(a) are easily mistaken for splice protectors

(b) have lower losses than fusion splices

(c) are quick and easy to fit

(d) are waterproof

4   An elastomeric splice:

(a) has losses which are about one thousandth of a fusion splice

(b) uses a flexible plastic to align the fibers

(c) is based upon the bent tube principle

(d) cannot be re-used

5   Some designs of mechanical splice can be easily mistaken for:

(a) an enclosure

(b) a PAS splicer

(c) a splice protector

(d) an all plastic fiber

12

Connectors

Connectors and adapters are the plugs and sockets of a fiber optic system. They allow the data to be rerouted and equipment to be connected to existing systems.

Connectors are inherently more difficult to design than mechanical splices. This is due to the added requirement of being able to be taken apart and replaced repeatedly. It is one thing to find a way to align two fibers but it is something altogether different if the fibers are to be disconnected and reconnected a thousand times and still need to perform well.

If two fibers are to be joined, each fiber has a connector attached and each is then plugged into an adapter. An adapter is basically a tube into which the two connectors are inserted. It holds them in alignment and the connectors are fixed onto the adapter to provide mechanical support. An adapter is shown as part of Figure 12.1.

Although different makes are sold as compatible, it is good practice to use the same manufacturer for the connector and for the adapter.

The design of connectors originated with the adaptation of those used for copper based coaxial cables and were usually fitted by the manufacturers onto a few meters of fiber called a *pigtail* which was then spliced into the main system.

Most connectors nowadays are fitted by the installer although pre-fitted ones are still available. The benefit of using the pre-fitted and pigtailed version is that it is much quicker and easier to fit a mechanical splice or perform a fusion splice than it is to fit a connector, so there is some merit in allowing the factory to fit the connector since this saves time and guarantees a high standard of workmanship.

When a connector is purchased, it always comes with a plastic dust cap to prevent damage to the polished end of the optic fiber. It is poor workmanship to leave fibers laying around without the caps fitted.

Before considering the details of the various types of connector, we will look at the main parameters met in their specifications so that we can make sense of manufacturers' data.

Connector parameters

Insertion loss

This is the most important measure of the performance of a connector. Imagine we have a length of fiber which is broken and reconnected by two connectors and an in-line adapter. If the loss of the system is measured and found to have increased by 0.4 dB then this is the value of the insertion loss. It is the loss caused by inserting a mated pair of connectors in a fiber. Be careful to ascertain whether the quoted loss for a connector is per mated pair or for each connector.

Typical value: 0.2–0.5 dB per mated pair.

Return loss

This is a measure of the Fresnel reflection. This power is being reflected off the connector back towards the light source. The lasers and LEDs used for multimode working are not greatly effected by the reflected power and so the return loss is not usually quoted in this instance. In singlemode systems the laser is effected and produces a noisy output. The laser suppliers will always be pleased to advise on permitted levels of return loss.

Typical value: –40 dB.

Mating durability

Also called Insertion loss change. It is a measure of how much the insertion loss is likely to increase in use after it has been connected and disconnected a large number of times.

Typical value: 0.2 dB per 1000 matings.

Operating temperature

These are, of course, compatible with the optic fiber cables.

Typical values: –25°C to +80°C.

Cable retention

Also called tensile strength or pull-out loading.

This is the loading that can be applied to the cable before the fiber is pulled out of the connector. It is similar in value to the installation tension on a light-weight cable.

Typical value: 200 N

Repeatability

This is a measure of how consistent the insertion loss is when a joint is disconnected and then remade. It is not a wear-out problem like mating durability but simply a test of whether the connector and adapter are designed so that the light path is identical each time they are joined.

This is an important feature of a connector but is not always quoted in specifications owing to the difficulty in agreeing a uniform method of measuring it. Some manufacturers do give a figure for it, some just use descriptive terms like 'high' or 'very high'. The quoted insertion loss should actually be the average insertion loss over a series of matings. Thus taking repeatability into account.

A survey of the main connectors

If we ignore the very early days when they were machined out of solid brass and had to be factory fitted, the first of the modern, fitted on site, connectors was the SMA (sub-miniature assembly). There are many similarities between the various types of connectors and this one serves as a good starting point. Connectors are nearly always assembled using epoxy resin and are not re-usable.

SMA (sub-miniature assembly – Figure 12.1)

The SMA connector has been superseded by the more modern designs but there are many still in use. To connect two fibers, we simply screw a connector onto

Figure 12.1

SMA connector

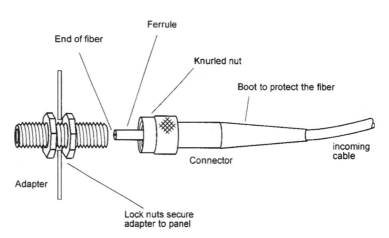

each end of the adapter. It is only used for multimode systems as the losses are too high for singlemode use.

The air gap

The length of the adapter ensures that the ends of the two ferrules are separated by an air gap small enough to allow the light to jump the gap to the other fiber.

This gives rise to the first problem. How tight do we tighten up on the screw thread? Not enough and the losses will be too high. Too tight and we will grind the end faces of the fiber together and will cause the glass to crack or be scratched. If this happens, the connector must be removed and thrown away.

Repeatability

The ferrules have a hole through the center to take the bare fiber (primary buffer stripped off) and are made of stainless steel or ceramic. In the case of the stainless steel, the 127 µm hole has to be drilled. If the hole is slightly off-center or over-size, it can cause eccentricity loss. Of the two, ceramic is by far superior. The ceramic is 'grown' on a piece of wire of precise thickness. When the wire is removed, we are left with a much more accurate hole.

Owing to misalignment and slight variations in the assembly of the connector there are losses which vary in magnitude as the connector is revolved within the adapter. This has the advantage of allowing us to optimize the connection by monitoring the losses as the connector is revolved. This is only OK if everyone who uses the connector has the test equipment, the time and patience to make the final adjustment. It is far better to have a known loss each time the connectors are mated by ensuring that it can only go together in a single fixed position. The SMA suffers from poor repeatability because it can be assembled in any random position. It is normal practice to insert the connector, measure the loss then revolve it through 90° and take a new measurement. This is done four times and the results are averaged.

Vibration

The simple screw thread offers very little protection against loosening when exposed to vibration.

The two versions, 905 and 906

The original SMA connector was called the SMA 905. To improve its performance the ferrule was modified and the new version was called the SMA 906 and is shown in Figure 12.2. One of its features was that the ferrule was stepped and to gain acceptability, a way was needed to make the 906 compatible with the previous version and the Delrin® sleeve was invented.

Figure 12.2

The two versions of the SMA

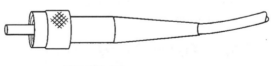

THE SMA 905

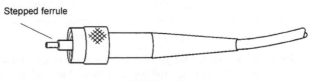

THE SMA 906

113

A Delrin sleeve is a piece of plastic sleeving that comes in two lengths, as seen in Figure 12.3. The half-length Delrin simply slips onto the 906 ferrule to bring

Figure 12.3

The two uses of a Delrin sleeve

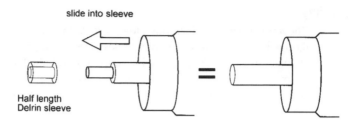

Converting a 906 ferrule into a 905

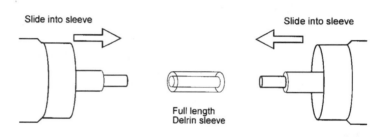

Aligning two 906s

its size up to the 905. This allows a 906 to be compatible with the earlier 905. The full length sleeve is a small plastic tube that slips over the end of the 906 ferrule and, within the adapter, the other 906 connector slots into the other end. The soft plastic ensures the two ferrules are held in accurate alignment rather like the elastomeric mechanical splice. Although only a few millimeters in length the improvement in the alignment is very significant. It was one of those really simple but amazingly effective 'why didn't I think of that?' type of inventions.

ST (straight tip — Figure 12.4)

This was developed by the US company, AT&T to overcome many of the problems associated with the SMA and is now the most popular choice of connector for multimode fibers. It is also available for singlemode systems.

The problem of repeatability is overcome by fitting a key to the connector and a corresponding keyway cut into the adapter. There is now only one position in which the connector can fit into the adapter.

The screw thread of the SMA has been replaced by a bayonet fitting so that there is no worry about the connector becoming loose when exposed to vibration.

The ferrule is spring loaded so that the pressure on the end of the ferrule is not

Figure 12.4

The ST connector

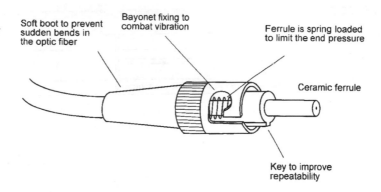

Soft boot to prevent sudden bends in the optic fiber

Bayonet fixing to combat vibration

Ferrule is spring loaded to limit the end pressure

Ceramic ferrule

Key to improve repeatability

under the control of the person fitting the connector. There are no SMA worries about how tight to do up the nut.

Polishing styles — Figure 12.5

The fiber through the center of the connector is polished during the assembly of the connector to improve the light transfer between connectors. There are three different styles called flat finish, physical contact (PC), and angled physical contact (APC). Many of the connectors are offered in different finishing styles so we see the connector name with a PC or APC added on the end. If nothing is mentioned, we assume a flat finish.

A flat finish is simply polished to produce a smooth flat end to the fiber so that the light comes straight out of the connector within the acceptance angle of the other fiber.

In the case of the PC finish, the fiber is polished to a smooth curve. There are two benefits of a PC connector. As the name implies, the two fibers make physical contact and therefore eliminates the air gap resulting in lower insertion losses. The curved end to the fiber also reduces the return loss by reflecting the light out of the fiber.

The APC finish results in very low return losses. It is simply a flat finish set at angle, typically 8°. The effect of this is that when the Fresnel reflection occurs much of the reflected power is at an angle less than the critical angle and is not propagated back along the fiber.

Fiber connector, physical contact (FCPC) — Figure 12.6

Also available as FC (flat finished) or FC APC (angled physical contact).

The FCPC is a high quality connector designed for long-haul singlemode systems and has very low losses. It can also be used for high quality multimode work if required and is often found on test equipment.

It looks like an SMA connector but it is keyed for repeatability.

The ferrule can be steel or ceramic inset in steel and is spring loaded or 'floating'. The end of the fiber is polished into the curved PC pattern. It can be polished flat if required and in this case it loses its PC suffix and just becomes an FC connector.

115

Figure 12.5

The alternative

finishes

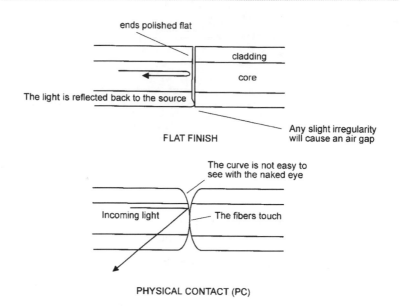

ends polished flat

cladding

core

The light is reflected back to the source

Any slight irregularity
will cause an air gap

FLAT FINISH

The curve is not easy to
see with the naked eye

Incoming light

The fibers touch

PHYSICAL CONTACT (PC)

Reduced Fresnel
reflection

Ends polished at an angle
(typically 8°)

ANGLED PHYSICAL CONTACT (APC)

Figure 12.6

The FCPC

connector

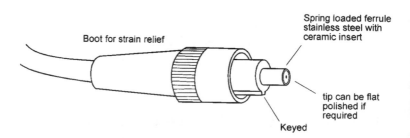

Spring loaded ferrule
stainless steel with
ceramic insert

Boot for strain relief

tip can be flat
polished if
required

Keyed

Mini-BNC — Figure 12.7

This was developed for the US market but has not proved popular and survives only because it is specified for the IBM token-ring network. Apart from being very slightly smaller, it has nothing to offer compared with the STPC.

In appearance, with its metal ferrule, it is easily mistaken for a BNC plug as used in copper based electrical systems.

It is for multimode use only, uses bayonet fittings and the ferrule is spring loaded and is a PC connector.

Figure 12.7

The mini-BNC

connector

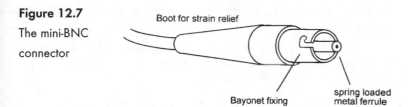

Boot for strain relief

Bayonet fixing

spring loaded
metal ferrule

Biconic connector

This is another connector, shown in Figure 12.8, which is also specified for the IBM token-ring network in its multimode form and is also widely used in the US for long haul singlemode telecommunications.

Figure 12.8

The biconic

connector

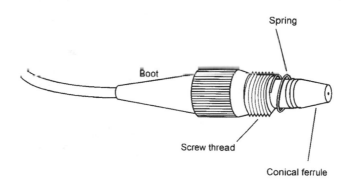

Spring

Boot

Screw thread

Conical ferrule

It is secured by screw thread rather than bayonet and has a spring loaded ferrule with a PC finish. When fitted to the adapter, the conical ferrule causes it to be self centering and thus providing low losses.

It is easily recognized by the unusual tapered ferrule and the exposed spring.

Subscriber connector (SC)

Also available in PC and APC versions and suitable for singlemode and multimode systems, it is illustrated in Figure 12.9.

Figure 12.9

The SC connector

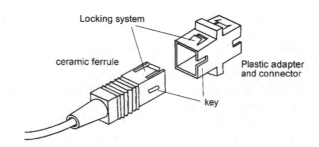

Locking system

ceramic ferrule

Plastic adapter
and connector

key

This connector is designed for high performance telecommunication and cable television networks. There is a different feel about this connector when compared

117

with the previous types. The body is of light plastic construction and has a more 'domestic' or 'office' feel about it. It has low losses and the small size and rectangular shape allows a high packing density in junction boxes. It plugs into the adapter with a very positive click action, telling us it's definitely engaged. A very nice connector destined to succeed.

Fixed shroud duplex (FSD) also called a media interface connector (MIC) — Figure 12.10

Unlike the other connectors, this one has two fibers within the same cover. This allows signals to be routed in two directions at the same time. This is called

Figure 12.10

The FSD (or MIC)

connector

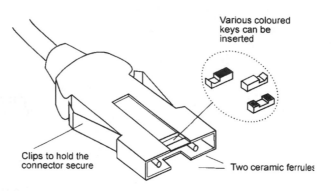

Various coloured keys can be inserted

Clips to hold the connector secure

Two ceramic ferrules

duplex operation.

It uses STPC ceramic ferrules, otherwise it is another all plastic connector, with a similar feel to the SC. It is intended to be used in local area networks (LANs) to interconnect computer systems and other pieces of office equipment.

Compared with the other connectors, it seems quite bulky and is designed to be easily handled and plugged into the equipment often by the end user rather than the system installer. It is keyed to prevent accidental insertion in the wrong socket and is color coded for easy recognition.

Adapters

Generally a system is designed to use the same type of connector throughout, and to ensure complete compatibility, and hence best performance, they are normally sourced from the same manufacturer.

Occasionally however, we meet two new problems.

The first is to connect two cables fitted with non-compatible connectors, say, an STPC connector to one fitted with a biconic connector. Such problems are easily solved by a wide range of 'something-to-something' type adapters. Some of these adapters do introduce a little extra insertion loss but not more than about 1 dB.

The second is to join a bare fiber to a system, quickly and easily, perhaps to connect a piece of test equipment or to try out a new light source. This is achieved by a bare fiber adapter. This is a misleading name since it is actually a connector as can be seen in Figure 12.11. It is really a bare fiber connector since it has to be plugged into an adapter. The only difference is that the fiber is held in position by a spring clip rather than by epoxy so that it can be readily re-used.

Figure 12.11
A bare fiber
adapter

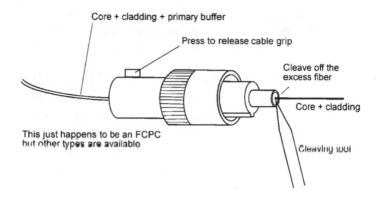

Core + cladding + primary buffer

Press to release cable grip

Cleave off the excess fiber

Core + cladding

This just happens to be an FCPC but other types are available

Cleaving tool

Fitting the bare fiber adapter

1   Strip off the primary buffer for about 25 mm or so and clean the fiber.

2   Press the cable grip and push in the fiber until it comes to a stop then release the cable grip. The primary buffer will not pass down the ferrule.

3   Cleave off the bit that sticks out of the end of the ferrule.

Done. Less than a minute.

The results depend on the quality of the cleave and are not as good as with a permanently fitted connector.

Plastic fiber connector — Figure 12.12

Plastic fiber connectors are very quick and easy to fit but the insertion loss is higher than normal for glass fibers — between 1 dB and 2 dB. The cables are

Figure 12.12
A connector for
plastic fibers

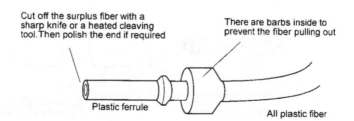

Cut off the surplus fiber with a sharp knife or a heated cleaving tool. Then polish the end if required

There are barbs inside to prevent the fiber pulling out

Plastic ferrule

All plastic fiber

connected in the usual method of having two connectors plugged into an in-line adapter. Sometimes the end of the plastic fiber is polished using a simplified version of the techniques used on glass fiber and sometimes it is cleaved off as in the bare fiber adapter.

Fitting a plastic fiber connector

1　The outer jacket (2.2 mm) is stripped off for about 25 mm.

2　The fiber is pushed into the connector as far as it will go.

3　The end is cleaved or polished according to the manufacturers instructions.

Note: there are barbs inside the connector to prevent the fiber pulling out in use. This also means that, once assembled, it is very difficult to get the connector off again so we need to study the instructions before the fiber is inserted.

Terminating a silica glass optic fiber (fitting the connector)

To avoid the job altogether, buy the connector already attached to a pigtail. Everything is done for us, all we have to do is to join it to the rest of the system by means of a fusion splice or mechanical splice.

The most usual method is called glue and polish. In essence, all that happens is that the fiber is stripped, glued into the connector and the end of the fiber is polished with abrasive film.

As usual, it is most important that we take some time out to read the instructions. It can save a lot of time and money in second attempts.

Fitting a connector on a silica fiber

1　Strip off the outer jacket, cut the Kevlar, and remove the primary buffer to the dimensions supplied with the connector (Figure 12.13).

Figure 12.13

Preparing the fiber

Crimp ring　　　　　　　　Boot

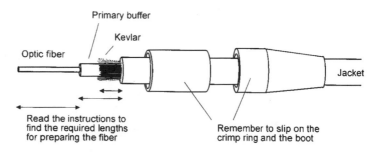

Primary buffer

Kevlar

Optic fiber

Jacket

Read the instructions to
find the required lengths
for preparing the fiber

Remember to slip on the
crimp ring and the boot

2　Slip the flexible boot and the crimp ring onto the fiber. The crimp ring is a metal tube about 10 mm in length which will grip the Kevlar and the connector to provide the mechanical support.

3   Clean the fiber with isopropyl alcohol in the way that was done prior to cleaving. Just as a practice run, carefully insert the stripped fiber into the rear of the ferrule and ease it through until the buffer prevents any further movement. If this proves difficult, it may help if the connector is twisted backwards and forwards slightly but be careful, the fiber must not break. Check that the fiber sticks out from the end of the ferrule. If it does break, the piece of fiber can be released with a 125 µm diameter cleaning wire which is available from suppliers.

4   Mix some two-part epoxy and load it into a syringe. The epoxy is often supplied in a sealed polyethylene bag with the hardener and adhesive separated by a sliding seal. Remove the seal and mix the adhesive and hardener by repeatedly squeezing the bag between the fingers. The mixing process can be aided by the use of a grooved roller which is rolled to and fro across the packet.

5   Insert the syringe into the connector until it meets the rear of the ferrule. Squeeze epoxy in slowly until a tiny bead is seen coming out of the front end of the ferrule. This shows that the ferrule is well coated with epoxy.

6   Carefully insert the stripped fiber into the rear of the ferrule until the buffer prevents any further movement. Again, take care not to break the fiber. If it does break, the fiber must be prepared again, as the dimensions will now be incorrect. The epoxy is very difficult to remove from the ferrule and the cleaning wire is not guaranteed to work under these conditions. Acetone may be helpful. This is a job best avoided.

7   Arrange the Kevlar over the spigot at the back end of the connector and slide the crimp ring over the Kevlar as shown in Figure 12.14. One end of the crimp ring should overlap the spigot and the other should cover the fiber jacket.

8   Using a hand crimping tool, crimp the Kevlar to the spigot and at the other end of the crimp ring, grip the cable jacket. This ensures that stress is taken by the Kevlar strength members and not by the optic fiber.

9   Put the connector into a small oven to set the epoxy. The oven, shown in Figure 12.15, is an electrically heated block of metal with holes to take the connectors. This will take about ten minutes at 80°C. When cured, the golden epoxy may have changed color to a mid to dark brown.

10   When it has cooled down, fit the boot.

11   Using a hand cleaver, gently stroke the fiber close to the end of the ferrule and lift off the end of the fiber (Figure 12.16). Store the broken end safely in a sealed receptacle for disposal.

12   The end of the fiber must now be polished. The easy way is to insert the fiber into a portable polishing machine and switch on.

Figure 12.14

Crimping

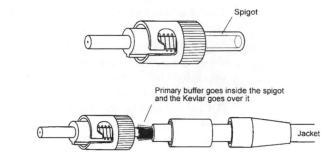

Spigot

Primary buffer goes inside the spigot
and the Kevlar goes over it

Jacket

Position the crimp ring over
the spigot and the jacket

Bare fiber

Crimp here to grip the Kevlar   and here to grip the jacket

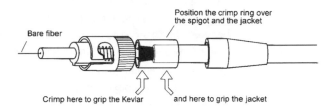

Figure 12.15

The connector goes
in the oven

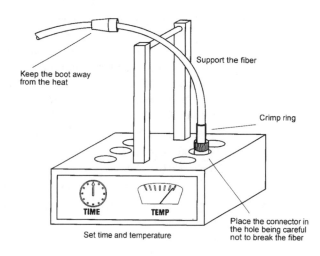

Keep the boot away
from the heat

Support the fiber

Crimp ring

TIME   TEMP

Set time and temperature

Place the connector in
the hole being careful
not to break the fiber

Figure 12.16

Cleave off the
surplus fiber

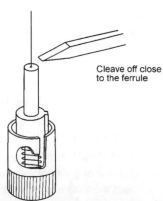

Cleave off close
to the ferrule

After about one minute, it's all over, the fiber is polished. The alternative is to do it ourselves.

13 Consult the instructions at this stage, each manufacturer has a recommended procedure for each type of connector and they usually know best. We will need a flat base of plate glass, hard rubber or foam about 200 mm square. A soft base is normally used for the PC finish.

The abrasive sheet is called a lapping film. It is a layer of aluminum oxide on a colored plastic sheet. Silicon carbide and diamond films are also available. The roughness of the abrasive is measured by the size of the particles and is colored to aid recognition. Grades vary from the coarse 30 µm colored green down to the ultra smooth white at 0.3 µm. Beware — not all types of film employ the same colors.

14 Using a magnifying glass, observe the end of the ferrule to see how much glass is protruding above the tiny head of epoxy (Figure 12.17). This unsupported glass is easily broken and

Figure 12.17

Carefully remove
the surplus fiber

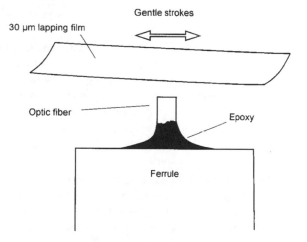

should be abraded down to the epoxy level by using a strip of coarse grade film (30 µm) held in the hand and stroked gently over the fiber. Be very careful not to apply too much pressure and stop when the epoxy is reached. If the fiber is too long or too much pressure is applied at this stage, the course lapping film will send shock waves down the fiber and it will crack. The crack has the characteristic 'Y' shape shown in Figure 12.18. It runs vertically down into the fiber and no amount of polishing will do anything to help the situation. We have lost a connector and gained some experience.

15 The fiber is supported perpendicular by a polishing tool called a dolly or a polishing guide (Figure 12.19). Each dolly is designed for a particular type of connector to ensure the correct

Figure 12.18

A characteristic

crack

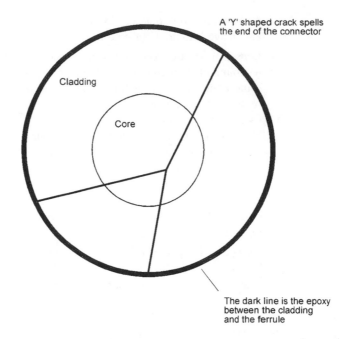

A 'Y' shaped crack spells
the end of the connector

Cladding

Core

The dark line is the epoxy
between the cladding
and the ferrule

dimensions and fitting mechanisms. The suppliers will always advise on the grades of film and methods to be used to be used. Once again it is worth reading the instructions carefully if an unfamiliar connector is being fitted. We start with the coarsest grade recommended, probably about 9 μm. Lay the film on the base material and attach the fiber in the dolly. Using only the weight of the dolly, slide the dolly in a figure of eight pattern for about eight circuits.

16 Using a microscope or magnifying eye glass, observe the end of the ferrule. There will be a large dark area which is the epoxy. If this is the case, repeat the above stage until the epoxy becomes lighter in color and eventually has a transparent feathered edge.

17 When this happens, remove the 9 μm film and clean the whole area, including the dolly and the connector. Clean it very carefully with a tissue moistened in alcohol or demineralised water. Make quite certain that no trace of abrasive is transferred from one stage to the next.

cleanliness is essential

18 Using a fine grade, about 3 μm, repeat the figures of eight. The end of the fiber takes on a bluish hue with some pale yellow from any remaining resin. Black marks are, as yet, unpolished areas or possibly water on the surface of the ferrule (dry it off before checking). A little more polishing and as the resin finally

disappears it is time to change grade of abrasive for the last time.

19   Another clean up, then a change of film to 1.0 μm or even 0.3 μm for the final polish. the end of the fiber will, hopefully, become clear and blue with no marks or scratches.

Some alternatives

Connectors are manufactured with hot-melt glue coating the inside as an alternative to using epoxy. This is convenient since we can always remelt it if we need to reposition the fiber.

Some epoxy resin is cured by ultraviolet light rather than by heat.

The polishing can be performed either dry or wet. Metal ferrules are often polished wet and ceramic dry, but there are too many exceptions to offer this as a rule. If wet polishing is recommended in the instructions, a few drops of wetting agent is applied to the surface of the lapping film. Not too much, otherwise the dully will tend to aquaplane over the surface without any polishing action. We should not use tap water as it includes impurities which under the microscope look like boulders and would scratch the surface of the fiber.

Final inspection

The finished connector should be inspected by a microscope with a magnification of at least 200 times. We are checking for surface scratches and chips caused by the polishing and also for cracks within the fiber rather than on the surface. Front lighting is good for spotting surface defects. Rear lighting, obtained by shining light down the fiber from the far end and therefore passing through the fiber is better for locating cracks within the fiber.

When observing any possible defects, the core is obviously of primary importance. The cladding is split into two halves. The outer part of the cladding does not greatly affect the operation of the fiber and we can be more forgiving of any failings in this area.

Main failing points — Figures 12.20 and 12.21

▷   Chips and cracks which extend into the inner part of the cladding.

▷   Cracks that extend for more than 25% of the circumference of the cladding.

▷   Scratches in the core area of a severity which is not consistent with the polishing techniques suggested by the manufacturer. This is what will happen if the grit from one stage in the polishing process is able to contaminate the finer grade lapping film.

Figure 12.20

Acceptable defects

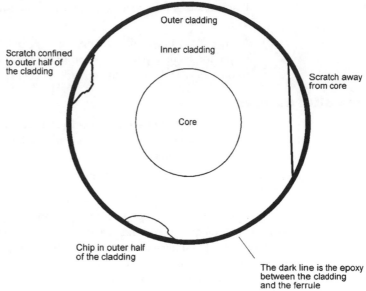

Outer cladding

Inner cladding

Scratch confined
to outer half of
the cladding

Scratch away
from core

Core

Chip in outer half
of the cladding

The dark line is the epoxy
between the cladding
and the ferrule

Figure 12.21

A total failure

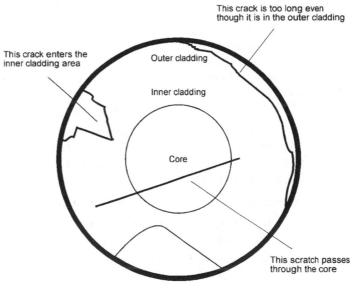

This crack is too long even
though it is in the outer cladding

This crack enters the
inner cladding area

Outer cladding

Inner cladding

Core

This scratch passes
through the core

Chip extends into inner half of cladding

Quiz time 12

In each case, choose the best option.

1   Two fibers can be joined by:

(a) two adapters plugged into each end of a connector

(b) a bare fiber connector

(c) two connectors and one adapter

(d) a single connector

2   A connector with a keyed ferrule and secured by a screw thread is likely to be an:

(a) SC

(b) FCPC

(c) STPC

(d) SMA 906

3   A PC finish:

(a) reduces both the return loss and the insertion loss

(b) makes physical contact but damages the end of the fiber

(c) is the result of polishing on a hard surface

(d) is the result of using an incorrect dolly

4   A full length Delrin sleeve:

(a) converts the ferrule of an SMA 905 connector to be compatible with the SMA 906

(b) can be used instead of an adapter

(c) is a popular choice of connector with cable television companies

(d) reduces the insertion loss of SMA 906 connectors

5   During polishing of a silica fiber, final inspection reveals a large scratch running right across the fiber. A likely cause of this is:

(a) contamination of the final lapping film with some coarse grit from a previous stage of the polishing

(b) using diamond lapping film instead of aluminum oxide film

(c) using the wrong dolly

(d) water laying on the surface of the fiber. Simply wipe it off with lint-free tissue

13

Couplers

Imagine an optic fiber carrying an input signal that needs to be connected to two different destinations. The signal needs to be split into two. This is easily achieved by a *coupler*. When used for this purpose, it is often referred to as a *splitter*.

Couplers are bi-directional, they can carry light in either direction. Therefore the coupler described above could equally well be used to combine the signals from two transmitters onto a single optic fiber. In this case, it is called a *combiner*. It is exactly the same device, it is just used differently.

The various ways of using couplers are shown in Figure 13.1.

Physically, they look almost the same as a mechanical splice, in fact in some cases we would need to count the number of fibers to differentiate between them. If there is one fiber at each end, it is a mechanical splice, any other number and it is a coupler.

Coupler sizes — Figure 13.2

A coupler with a single fiber at one end and two at the other end would be referred to as a 1 x 2 coupler (read as *one by two*). Although 1 x 2, and 2 x 2, are the most common sizes they can be obtained in a wide range of types up to 32 x 32 and can be interconnected to obtain non-standard sizes. Splitters are more common than combiners and this has made it more natural to refer to the single fiber end as the input.

The numbering of the ports is shown in Figure 13.3 (*port* is just a fancy word used in electronics to mean a connection).

Splitting ratio or coupling ratio

The proportion of the input power at each output is called the splitting ratio or coupling ratio. In a 1 x 2 coupler, the input signal can be split between the two

Figure 13.1

The uses of

couplers

A COUPLER CAN CONNECT.....

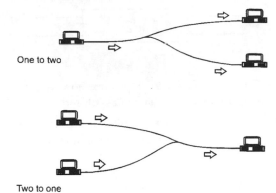

One to two

Two to one

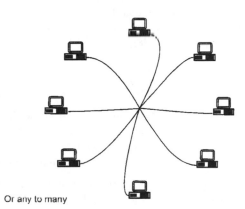

Or any to many

Figure 13.2

The most common

types of coupler

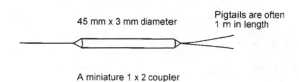

45 mm x 3 mm diameter

Pigtails are often
1 m in length

A miniature 1 x 2 coupler

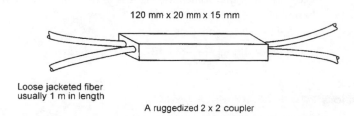

120 mm x 20 mm x 15 mm

Loose jacketed fiber
usually 1 m in length

A ruggedized 2 x 2 coupler

Figure 13.3
One possible
numbering system
for the ports

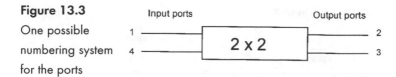

outputs in any desired ratio. In practice however, the common ones are 90:10 and 50:50. These are also written as 9:1 and 1:1.

In the cases where the splitting ratio is not 1:1, the port which carries the higher power is sometimes called the *throughput port* and the other is called the *tap port*.

Coupling tolerance

Even when the splitting ratio is quoted as 1:1, it is very unlikely, due to manufacturing tolerances that the input power is actually shared equally between the two outputs. The acceptable error of between 1% and 5% is called the coupling or splitting tolerance.

Losses

A gloomy note before we start

When consulting trade publications, we find that the terms used to describe coupler losses, the naming of the ports and even the numbering of the connections have not been totally standardized. This makes it difficult to avoid meeting several different versions of the formula for each loss.

The only way to combat this is to understand the nature of the losses and then to be fairly flexible when it comes to the way it has been expressed.

Referring to Figure 13.4, the losses are stated in decibels and assume that the input is applied to port 1 and the output is taken from ports 2 and 3. For the moment, we will ignore the other connection shown as port 4 with its outward pointing arrow. This will be discussed further when we look at directionality loss.

Figure 13.4
Assumed power
directions for
definitions

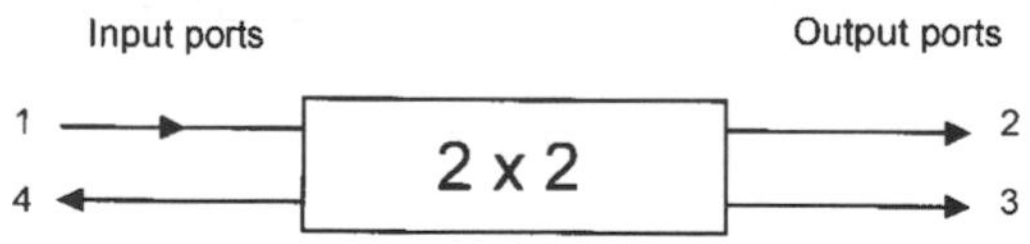

Input ports   Output ports

1 →   → 2
      2 x 2
4 ←   → 3

We may recall that, generally, the loss in decibels is derived from the standard formula:

$$\text{Loss} = 10\log\left(\frac{\text{power}_{\text{out}}}{\text{power}_{\text{in}}}\right) \text{ dB}$$

Excess loss

Excess loss is a real loss. If 10 mW goes into a device and only 9 mW comes out, then it is reasonable enough to think of the other 1 mW to be a loss. The light energy has been scattered or absorbed within the coupler and is not available at the output. So what we are really saying is that the loss is dependent on the total output power compared to the input power. In the case of the coupler in Figure 13.4, the output power is the sum of ports 2 and 3 and the input is at port 1.

So excess loss would look like this:

$$\text{Excess loss} = 10\log\left(\frac{P_2 + P_3}{P_1}\right) \text{ dB}$$

where P_1, P_2, P_3 are the power levels at the respective ports.

Directionality loss or crosstalk or directivity

When we apply power to port 1 we expect it to come out of ports 2 and 3 but not out of port 4, the other input port. Unfortunately, owing to backscatter within the coupler, some of the energy is reflected back and appears at port 4. This backscatter is very slight and is called directionality loss or crosstalk. The fact that the backscatter comes out of port 4 accounts for the direction of the arrow in Figure 13.4.

$$\text{Directionality loss} = 10\log\left(\frac{P_4}{P_1}\right) \text{ dB}$$

A typical figure is –40 dB.

Directivity puts the same information around the other way, if the reflected power has a level of –40 dB, then the power which is *not* reflected has a ratio of +40 dB. In the formula, the power levels are just inverted.

$$\text{Directivity} = 10\log\left(\frac{P_1}{P_4}\right) \text{ dB}$$

Insertion loss or port-to-port loss or throughput loss or tap loss

This looks at a single output power compared with the input power, so in Figure 13.4 there are two possibilities. We could look at the power coming out of port 2 and compare it with the input power at port 1 or we could do a similar thing with port 3 compared with the input power at port 1.

Generally, insertion loss for any output port could be written as:

$$\text{Insertion loss} = 10\log\left(\frac{P_{\text{output port}}}{P_{\text{input port}}}\right) \text{ dB}$$

As an example, the insertion loss at port 2 is:

$$\text{Insertion loss} = 10\log\left(\frac{P_2}{P_1}\right) \text{ dB}$$

This would then be referred to as the insertion loss of port 2 or simply the *port-to-port loss* between ports 1 to port 2.

If, in the above example, the splitting ratio was not 1:1, then port 2 may be referred to as the throughput port and so the formula above becomes the throughput loss. Similarly, if ports 3 and 1 were used, the loss could be called the tap loss.

Coupling loss

This is often overlooked. Whenever a coupler is used, it has to be joined to the rest of the circuit. This involves the use of connectors or splices. The losses caused by these connectors or splices must be added to the losses introduced by the coupler.

Example

Calculate the output power at each port in the coupler shown in Figure 13.5.

Figure 13.5

A worked example

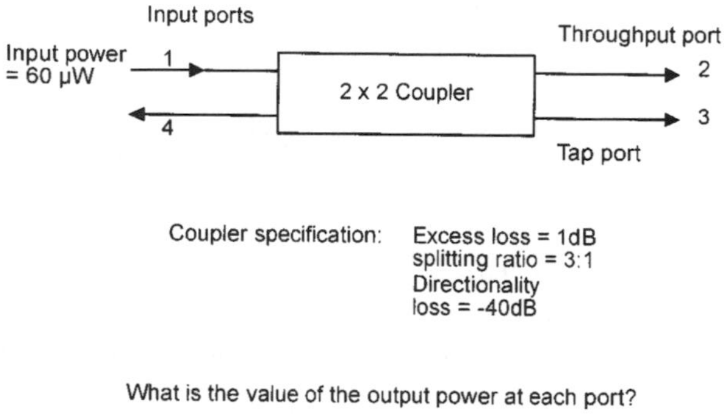

Coupler specification: Excess loss = 1dB
splitting ratio = 3:1
Directionality
loss = -40dB

What is the value of the output power at each port?

Output power at port 4

The directionality loss is quoted as –40 dB.

Starting with the standard formula for decibels.

$$\text{Loss} = 10\log\left(\frac{\text{power}_{\text{out}}}{\text{power}_{\text{in}}}\right) \text{ dB}$$

The input power is 60 µW and the loss figure in decibels is –40 dB so we make a start by inserting these figures into the formula:

$$-40 \text{ dB} = 10\log\left(\frac{\text{power}_{\text{out}}}{60 \times 10^{-6}}\right)$$

Divide both sides by 10:

$$-4 = \log\left(\frac{\text{power}_{\text{out}}}{60 \times 10^{-6}}\right)$$

Take the antilog of each side:

$$10^{-4} = \left(\frac{\text{power}_{out}}{60 \times 10^{-6}}\right)$$

So:

$$60 \times 10^{-6} \times 10^{-4} = \text{power out}$$

So power out of port 4 = 60×10^{-10} = 600 nW.

As the output from port 4 is so small, it is often ignored.

Output power at port 2

Port 2 is the throughput port i.e. the port with the largest output power. With a splitting ratio of 3:1, for every 4 units of power leaving the coupler there are three at port 2 and only 1 at port 3. This means that 0.75 of the power leaving the coupler goes via port 2.

But how much power is leaving the coupler? This is the input power minus the excess loss. Port 4 output can be ignored since it is so slight compared with the other power levels.

In this example, the excess loss is 1 dB so if we convert this 1 dB into a ratio, we can find the output power.

Into the standard formula we put the -1 dB (minus, as it is a loss) and the input power:

$$-1\ \text{dB} = 10\log\left(\frac{\text{power}_{out}}{60 \times 10^{-6}}\right)$$

Divide both sides by 10:

$$-0.1 = \log\left(\frac{\text{power}_{out}}{60 \times 10^{-6}}\right)$$

Take the antilog of each side:

$$10^{-0.1} = \left(\frac{\text{power}_{out}}{60 \times 10^{-6}}\right)$$

So:

$$60 \times 10^{-6} \times 10^{-0.1} = \text{power out}$$

By calculator, $10^{-0.1} = 0.794$

So total power out of the coupler = input power $\times$ 0.794, or:

$$60 \times 10^{-6} \times 0.794 = 47.64 \times 10^{-6}\ \text{W}$$

Of the 60 µW that entered the coupler, 47.64 µW is able to leave. Of this amount 0.75 leaves via port 2 so:

$$\text{power out of port 2} = 0.75 \times 47.64\ \mu\text{W} = 35.73\ \mu\text{W}$$

Output power at port 3

We have already calculated the power remaining after the excess loss to be 47.64 µW and since we are dealing with port 3, the tap port, the proportion of power leaving by this port is only 0.25 of the total.

Thus the output power at port 3 is 0.25 × 47.64 µW = 11.91 µW.

The results are shown in Figure 13.6.

Figure 13.6

The resulting power levels

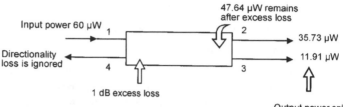

Input power 60 µW

47.64 µW remains after excess loss

35.73 µW

11.91 µW

Directionality loss is ignored

1 dB excess loss

Output power split in a ratio of 3:1

The tee-coupler — Figure 13.7

This is simply a 1 × 2 coupler used to convey a single signal to a number of different work stations. Such stations are said to be connected on a network. It would use a high splitting ratio of 9:1 or similar to avoid draining the power from the incoming signal.

Advantages and disadvantages of a tee network

The main advantage is its simplicity. The couplers are readily available and, if required, can be supplied with connectors already fitted. This means that the network can be on-line very quickly indeed.

The disadvantage is the rapid reduction in the power available to each of the workstations as we connect more and more terminals to the network. As the power is reduced, the number of data errors increases and the output becomes increasingly unreliable. At first glance we could solve this problem by simply increasing the input power level. However we run the very real risk of overloading the first workstation.

Power levels in a tee network

Specification for our example system:

> ▷ incoming power = 1mW

> ▷ splitting ratio of each coupler = 9:1

> ▷ excess loss of each coupler = 0.3 dB. The couplers are joined by connectors with an insertion loss of 0.2 dB each.

Step 1 — Figure 13.8

The incoming power level is reduced by 0.2 dB by the first connector, and 0.3 dB by the excess loss.

Total power reduction is 0.2 + 0.3 = 0.5 dB.

By inserting the values into the standard decibel formula, remembering to use

Figure 13.7

A tee network

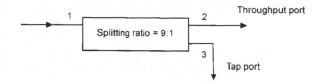

Figure 13.8

The first terminal

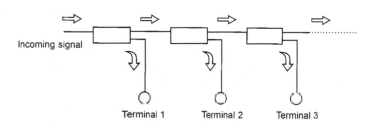

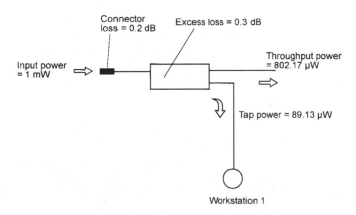

–0.5 dB as it is a loss, we have:

$$-0.5 \text{ dB} = 10\log\left(\frac{\text{power}_{\text{out}}}{1 \times 10^{-3}}\right)$$

Divide both sides by 10:

$$-0.05 \text{ dB} = \log\left(\frac{\text{power}_{\text{out}}}{1 \times 10^{-3}}\right)$$

And antilog:

$$0.8913 = \left(\frac{\text{power}_{\text{out}}}{1 \times 10^{-3}}\right)$$

So, the input power is:

$$0.8913 \times 1 \times 10^{-3} = \text{power out} = 891.3 \ \mu\text{W}$$

135

Step 2

The 891.3 µW is the power just before it is split into the two output ports.

As the splitting ratio is 9:1, the throughput power at port 2 is 0.9 of the available power or 802.17 µW. Similarly, the tap power is 0.1 of 891.3 µW or 89.13 µW.

Step 3 — Figure 13.9

The throughput power, 802.17 µW is actually the input power to the next section of the network and is simply a replacement for the 1 mW input in Step 1.

This new input power suffers the same connector insertion loss, coupler excess loss and splitting ratio and so the calculations would involve exactly the same steps as we have already used.

The results we would obtain are throughput loss = 643.47 µW and the tap power going to terminal 2 = 71.49 µW.

Step 4

The next section would decrease the powers by the same proportions and it would result in a throughput loss of 516.2 µW and a tap power of 57.4 µW.

The same proportional loss would occur at each section of the network.

The star coupler

This is an alternative to the tee coupler when a larger number of terminals is involved as shown in Figure 13.10.

The star coupler takes the input signal to a central location, then splits it into many outputs in a single coupler. Styles of up to 1 × 32 and up to 32 × 32 are commonly available.

Advantages and disadvantages

The main advantage of using star couplers is that the losses are lower than a tee coupler for networks of more than three or four terminals as in Figure 13.11. This is because the star coupler requires only one input connector and suffers only one excess loss. The larger the number of terminals, the more significant are the benefits.

The disadvantage is that the star coupler will normally use much larger quantities of cable to connect the terminals since the star is located centrally and a separate cable is connected to each of the terminals. A tee network can use one cable to snake around the system from terminal to terminal.

Networks

Couplers can be employed to produce *local area networks* (LANs). This is an interconnection of terminals called *nodes* over a limited area such as a university campus or a business park. The general layout patterns are called

Figure 13.9

The power levels

quickly decrease

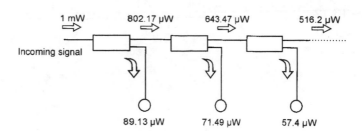

Figure 13.10

Two forms of 4 x 4

star couplers

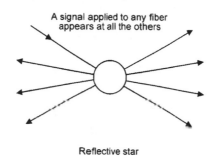

A signal applied to any fiber
appears at all the others

Reflective star

A signal applied to any of the inputs
1 - 4, appears at all of the outputs

Input 1
Input 2
Input 3
Input 4

Output 1
Output 2
Output 3
Output 4

Transmissive star

Figure 13.11

Loss comparison

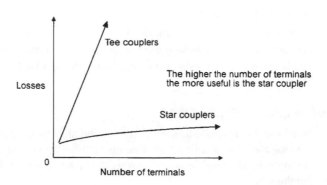

Tee couplers

The higher the number of terminals
the more useful is the star coupler

Star couplers

Losses

0

Number of terminals

network topologies and are illustrated in Figure 13.12.

Larger networks spread over a wider area are referred to as *wide area networks*

Figure 13.12
Some ways of
connecting a
network

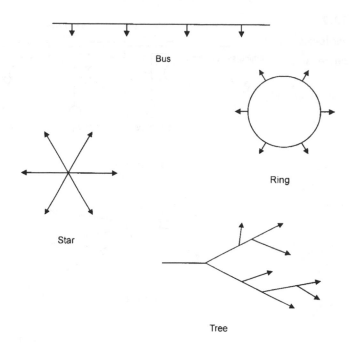

Bus

Ring

Star

Tree

(WANs), and those restricted to a particular city or town are called *metropolitan area networks* (MANs).

Construction of couplers

Fused couplers

This is the most popular method of manufacturing a coupler. It is, or appears to be, a very simple process.

The fibers are brought together and are then fused just like in a fusion splicer as seen in Figure 13.13. The incoming light effectively meets a thicker section of fiber and spreads out. At the far end of the fused area, the light enters into each of the outgoing fibers.

A fused star coupler is made in a similar way (Figure 13.14). The fibers are twisted round to hold them in tight proximity, then the center section is fused. In the case of the reflective star, the fibers are bent back on themselves before being fused.

Mixing rod couplers — Figure 13.15

If several fibers are connected to a short length of large diameter fiber, called a mixing rod, the incoming light spreads out until it occupies the whole diameter of the fiber. If several fibers are connected to the far end they each receive some of the light.

A reflective coupler can be produced by putting a mirror at the end of the mixing rod. The light traveling along the mixing rod is reflected from the end mirror and all the attached fibers receive an equal share of the incoming light.

Figure 13.13

Light sharing in a
fused coupler

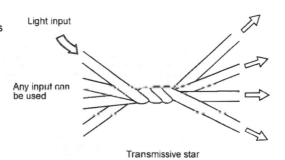

Figure 13.14

Fused star couplers

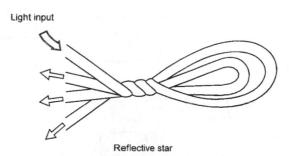

Variable coupler — Figure 13.16

This is more of an experimental or test laboratory tool than for the installation environment. It enables the splitting ratio to be adjusted to any precise value up to 19:1, which allows us to try out the options before the final type of coupler is purchased.

The design principle is very simple. A vernier adjustment allows precise positioning of the incoming fiber so that the light can be split accurately between the two output fibers to provide any required splitting ratio.

This form of variable coupler is available for all plastic as well as glass fibers, singlemode and multimode.

Figure 13.15

Star couplers by mixing rods

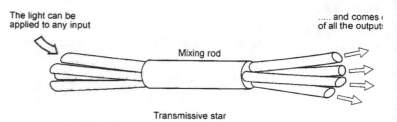

The light can be applied to any input

..... and comes of all the output

Mixing rod

Transmissive star

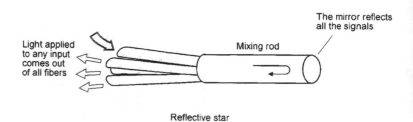

The mirror reflects all the signals

Light applied to any input comes out of all fibers

Mixing rod

Reflective star

Figure 13.16

The variable coupler

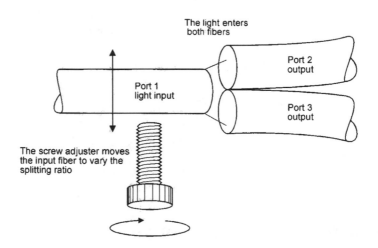

The light enters both fibers

Port 2 output

Port 1 light input

Port 3 output

The screw adjuster moves the input fiber to vary the splitting ratio

Quiz time 13

In each case, choose the best option.

1   A 4 × 4 coupler would have a total of:

(a) 16 ports

(b) 4 ports

(c) 9 ports

(d) 8 ports

2   In the coupler shown in Figure 13.17, the tap power would have a value of approximately:

(a) 18.2 μW

(b) 0.182 W

(c) 163.8 μW

(d) 21.9 μW

3   The main advantage of using a star coupler to connect a large number of terminals is that:

(a) more cable would be used and hence system reliability would increase

(b) the power loss is lower than would be the case if tee couplers were to be used

(c) higher levels of data errors could be tolerated

(d) less cable is used

4   Coupling ratio is also known as:

(a) directionality loss

(b) coupling loss

(c) splitting ratio

(d) directivity ratio

5   The output connection which carries the highest power level is the:

(a) tap port

(b) power port

(c) star

(d) throughput port

Figure 13.17

Question 2

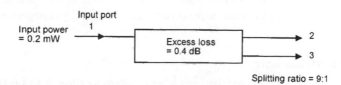

Splitting ratio = 9:1

14

Light sources and detectors

Most light sources and detectors are electronic devices built from the same semiconductor materials as are used in transistors and integrated circuits.

The design of these devices is a separate study and will not be considered here. Instead, our view will be restricted to the characteristics which are of interest to the user.

Lasers

The most common form of laser diode is called an *injection laser diode* (ILD) or just *injection diode* (ID). The word *injection* is not of interest — it merely refers to part of the process occurring inside the semiconductor material.

A laser provides a light of fixed wavelength which can be in the visible region around 635 nm or in any one of the three infrared windows. The light has a very narrow bandwidth, typically only a few nanometers wide. This ensures that chromatic dispersion is kept to a low value and this, together with fast switching, allows high data transmission rates.

As the laser device itself is barely visible to the unaided eye, it must be contained in some form of package. Two typical examples are shown in Figure 14.1.

Lasers for visible light

The light is launched via a lens system to allow it to be concentrated into a beam. Visible laser light finds applications in bar code readers, CD players,

Figure 14.1

Lasers

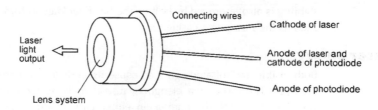

Laser light output

Connecting wires

Cathode of laser

Anode of laser and cathode of photodiode

Anode of photodiode

Lens system

A typical package for visible light lasers and for 850 nm lasers

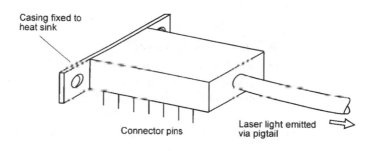

Casing fixed to heat sink

Connector pins

Laser light emitted via pigtail

A laser for singlemode use

medical and communication systems. They are usually fitted with a built-in light detector so that they can receive reflected information as in the case of the bar code reader.

Lasers for 850 nm use

These can be packaged in either of the ways illustrated in Figure 14.1 depending on their application. The fact that their output is not visible allows for use in security, ranging, automotive and industrial and military applications. They also provide the light source for short and medium range fiber communications.

Lasers for singlemode communications

Successful launching into singlemode fibers requires very high precision and this is achieved by optimizing the position of an attached pigtail which can then be connected to the main fiber run by any desired method.

A photoelectric cell is also included as a monitoring device to measure the output power. This provides feedback to allow for automatic control of the laser output power.

The output power of a laser is affected by any change in its temperature, generally decreasing in power as the temperature increases. Some laser modules include a temperature sensor to combat this problem. It provides internal temperature information which is used to control a thermo-electric cooler like a small refrigerator, to maintain the temperature. The temperature

143

stability is also improved by bolting the laser package to some form of heat sink such as the instrument casing.

Laser safety

Both visible and infrared light can cause immediate and permanent damage to the eyes. The shorter wavelengths cause damage to the retina and the longer wavelengths attack the cornea, in either case medical science can offer no remedy once the damage is done. Permanent loss of eyesight in less than a second by exposure to light we can't even see — it doesn't seem fair somehow.

It is extremely important that we take sensible precautions.

Never look into:

> ▷   a live laser source

> ▷   an unknown light source

> ▷   any fiber until you have ascertained that it is safe. Check it yourself even if trusted colleagues say 'its OK we've just checked it out'. They may be talking about a different fiber or they may have made a mistake.

If an instrument such as a live fiber detector is used, make sure it is working.

Beware of concentrating the light by instruments such as will happen when checking a cleave or the end condition of a connector with a microscope.

Laser classifications are based on an international standard titled *Radiation Safety of Laser Products, Equipment Classification, Requirements and User's Guide,* referred to as *IEC standard 825.* Additional national standards apply in each country.

The IEC 825 classification has used four classes of laser based on the accessible emission limit (AEL). Every laser must carry a warning label stating the class of laser as shown in Figure 14.2. It is the responsibility of the manufacturer to determine the classification of the laser and they do so by measuring the wavelength, output power and the pulsing characteristics.

IEC classifications

Class 1

Safe under reasonably foreseeable conditions of operations. Note that it doesn't say 'safe under any conditions'.

Class 2

Visible lasers with light output within the visible spectrum of 400–700 nm. There is an assumption here that the blink reflex will close the eyes within a fraction of a second and hence provide protection. Prolonged exposure will cause damage.

Class 3a

Safe for viewing by the unaided eye either visible or infrared light but possibly unsafe when viewed with instruments.

Figure 14.2
Laser warning
labels

HAZARD LABEL

CAUTION
INVISIBLE LASER RADIATION
CLASS 3A

Power .... Wavelength ....

EXPLANATORY LABEL

Class 3b

Direct viewing is hazardous but reflected light is normally OK. Note the *normally*. Not to be viewed with instruments.

Class 4

Horribly dangerous. Even reflections are hazardous and the direct beam can cause fires and skin injury.

Control measures

For classes 2, 3 and 4, control measures are employed such as interlocks, keys, laser 'on' warning lights, remote switching, prevention of reflections across walkways. The precautions depend on the situation, use and power of the laser. The appropriate national standards as well as IEC 825 should be consulted for guidance.

Laser specifications

Wavelength

The wavelength quoted is only a typical value. So if we want to buy a laser for the 1300 nm window, the one offered may well be quoted as 1285–1320 nm and the actual frequency will fall somewhere between these limits. Sometimes it would just be sold as 1300 nm (nominal).

145

Rise and fall time — Figure 14.3

This is a measure of how quickly the laser can be switched on or off measured between the output levels of 10% to 90% of the maximum. A typical value is 0.3 ns.

Threshold current — Figure 14.4

This is the lowest current at which the laser operates. A typical value is 50 mA and the normal operating current would be around 70 mA.

Spectral width — Figure 14.5

This is the bandwidth of the emitted light. Typical spectral widths lie between 1 nm and 5 nm. A laser with an output of 1310 nm with a spectral width of 4 nm, would emit infrared light between 1308 nm and 1312 nm.

Operating temperature

No surprises here. Typical values are −10°C to +65°C and therefore match the temperature ranges of fibers quite well.

Voltages and currents

The specifications also list the operating voltages and currents of the monitor detector, the cooler current and the thermistor resistance. These are generally only of interest to the equipment designer or the repair technician.

Output power

The output power may be quoted in watts or in dBm.

LEDs — light emitting diodes

LEDs can provide light output in the visible spectrum as well as in the 850 nm, 1350 nm and the 1500 nm windows.

Compared with the laser, the LED has a lower output power, slower switching speed and greater spectral width, hence more dispersion. These deficiencies make it inferior for use with high speed data links and telecommunications. However it is widely used for short and medium range systems using both glass and plastic fibers because it is simple, cheap, reliable and is less temperature dependent. It is also unaffected by incoming light energy from Fresnel reflections etc.

Although the lower power makes it safer to use, it can still be dangerous when the light is concentrated through a viewing instrument. Typical packages are shown in Figure 14.6.

PIN diodes

A PIN diode is the most popular method of converting the received light into an electronic signal. Their appearance is almost identical to LEDs and lasers. Indeed the diagrams in Figure 14.6 would serve equally well for PIN diodes if the labels were changed. They can be terminated with SMA, ST, SC, biconic

Figure 14.3

Rise & fall time

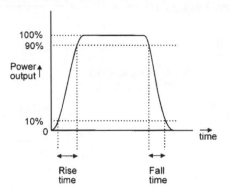

Figure 14.4

Laser characteristic

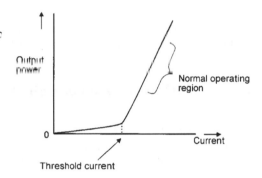

Figure 14.5

Spectral widths of
Leds and lasers

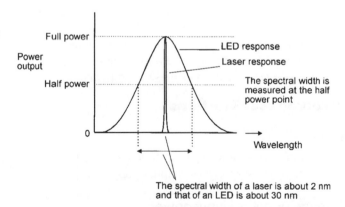

and a variety of other connectors or a pigtail.

It may be of interest to have a brief look at its name. It uses a semiconductor material, either germanium or silicon. The pure semiconductor material is called an intrinsic semiconductor — this is the *I* in the name. To make it work, we have to add a controlled amount of impurity into the semiconductor to change its characteristics. The semiconductor is converted into two types, one called P-type semiconductor and the other called N-type. These are arranged either side of the *I* material to make an *I* sandwich. Hence P–I–N or PIN diode. The theory of its operation will not be considered further.

Figure 14.6

LEDs —

encapsulated and

SMA packages

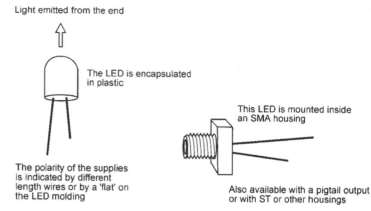

Light emitted from the end

The LED is encapsulated in plastic

The polarity of the supplies is indicated by different length wires or by a 'flat' on the LED molding

This LED is mounted inside an SMA housing

Also available with a pigtail output or with ST or other housings

While we can still buy straightforward PIN diodes, it is more usual for it to have an amplifier built into the module to provide a higher output signal level.

Avalanche diode also called an avalanche photo diode or APD

Higher output signals can be achieved by an avalanche diode. It uses a small internal current to generate a larger one in the same way that a snow-ball rolling down a mountainside can dislodge some more snow which, in turn, dislodges even more snow and eventually gives rise to an avalanche.

They have the advantages of a good output at low light levels and a wide dynamic range — it can handle high and low light levels. However there are a number of disadvantages which tend to outweigh the benefits. It has higher noise levels, costs more, generally requires higher operating voltages and its gain decreases with an increase in temperature.

Light receiver specifications

Wavelength

This is quoted as a range e.g. 1000 nm to 1600 nm, or by stating the frequency that provides the highest output e.g. peak wavelength = 850 nm.

Dynamic range or optical input power

Dynamic range is the ratio of the maximum input power to the lowest. It is quoted in decibels e.g. 21 dB.

The optical input power is the same information expressed in watts. e.g. 1 μW to 125 μW.

Responsivity

A measure of how much output current is obtained for each watt of input light. e.g. $0.8AW^{-1}$. This means that the current will increase by 0.8 amps for every watt of increased light power.

Response time

This is the rise and fall time that we saw in Figure 14.3. It determines the fastest switching speed of the detector and hence limits the maximum transmission rate e.g. t_r or $t_f = 3.5$ ns.

Bit rate or data rate or bandwidth

These are both measures of the maximum speed of response to incoming signals and is therefore determined by the response time above.

Quiz time 14

In each case, choose the best option.

1   The safest type of laser is referred to as:

(a) infrared

(b) class 1

(c) pulsed

(d) class 4

2   A typical value for the spectral width of a laser is:

(a) 1310 nm

(b) 3 nm

(c) 850 nm

(d) 30 nm

3   An APD:

(a) can produce visible light as well as infrared light at 850 nm, 1300 nm and 1550 nm

(b) has good electrical output in low light conditions

(c) has a lower dynamic range than a PIN diode

(d) is cheaper than a PIN diode

4   LEDs are not used as the light source for high speed telecommunications because of their:

(a) higher cost

(b) poor reliability

(c) inability to provide a visible light output

(d) slow switching speed and higher spectral width

5   A typical value of t_r for a laser would be:

(a) 3.5 ns

(b) 3 nm

(c) 90%

(d) 0.3 ns

15

Testing a system

Does it work? This is the first question that occurs to us once we have carried out any work on an optic fiber system. There are several different tests we can conduct. We will start with the simplest.

Remember to take precautions to avoid accidental viewing of 'live' fibers
---

Visible light continuity test

For short lengths of cable up to about 100 meters any source of white light can be used. The end of the fiber can be held against a flashlight or an electric light bulb or even sunlight if it is available. The range obviously depends on the brightness of the light source and the losses on the fiber so an absolute limit is impossible to quote. This method is commonly used to test short connecting short cables called patchcords, patch cables or jumper cables. These are short lengths of easily replaceable cable, typically 5–20 meters in length, used to connect between enclosures and instruments. At the far end, our eyes can detect the light more easily if we switch it on and off rather than just leaving it permanently on. The range can be extended up to a kilometer using an ultrabright red light LED.

Light source and power meter

The previous methods have provided a quick and easy go/no go test over a limited distance. The benefit of using a light source and power meter is that they are able to measure the actual power loss of the fiber system.

Light source — Figure 15.1

A light source is a hand-held instrument able to provide a light output within one or more of the standard windows: visible, 850 nm, 1300 nm and 1550 nm

Figure 15.1

A light source

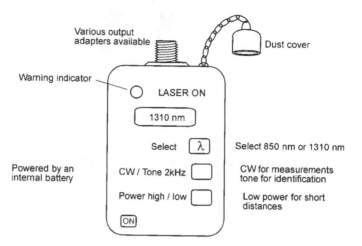

Various output adapters available

Dust cover

Warning indicator

LASER ON

1310 nm

Select   λ   Select 850 nm or 1310 nm

Powered by an internal battery

CW / Tone 2kHz   CW for measurements tone for identification

Power high / low   Low power for short distances

ON

using an LED or laser light source. They often provide outputs at more than one of the wavelengths, as installation contracts generally require measurements to be taken at two different wavelengths. A popular choice being 850 nm and 1300 nm.

For reliable results, the power output of the light must be very stable over the period of the test, typically within 0.1 dB over 1 hour. The output is able to be switched between a test tone of 2 kHz or 270 Hz or 10 kHz and a continuous output called CW (continuous wave). The choice of test tones allow easy identification of fibers under test.

Power meter

At first glance, these look very similar to the light sources. Compare Figures 15.1 and 15.2. They are often sold as matching pairs though there is nothing to prevent any light meter and power meter to be used together provided they are compatible.

The wavelength is adjustable over the three windows and some offer a facility to step up and down by small increments. This allows the fiber characteristics to be quoted at any required wavelength. It is a 'nice to have' rather than an essential feature.

The power levels can be indicated in µW or in decibels as dBm, relative to one milliwatt or as dBr, relative to a previously noted value.

They are available with internal memories to store the day's work and a thermal printer for hard copies.

Calibration (or traceability)

If the light source and power meter are to be used to check an installation or repair on a commercial basis, the customer will need assurance that your

Figure 15.2

Power meter

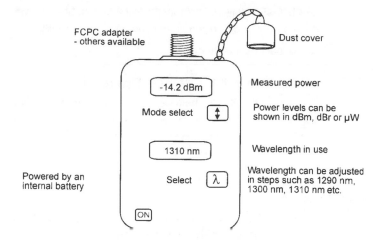

instruments are telling the truth. The proof of this is provided by a calibration certificate for each instrument which must be renewed at intervals, usually annually. The calibration must be carried out by an authorized company whose instruments themselves are calibrated against the appropriate national standards. In this way we can trace the accuracy back to its source.

An example measurement

To measure the losses in a typical link as shown in Figure 15.3, we start by selecting a patchcord about 20 m in length. If possible, it should be constructed from fiber of the same type and fitted with identical connectors to those used in the link to be tested. In cases where this is not possible, reduced accuracy in the final result will have to be tolerated. Keep a record of the patchcord and the instruments used so that the measurement could be repeated if a doubt arises at a later date.

Figure 15.3

The optic fiber to
be tested

Step 1 Setting up the light source and power meter

Connect the meters together using the patchcord as shown in Figure 15.4. Switch them on. Select the required wavelength and, on the power meter, switch to dBm mode.

Wait until the readings have stabilized. At this stage, the power meter will indicate the incoming power level in dBm. Set the power meter to dBr and it will accept the incoming light level as the reference level. The reading will now change to 0 dBr.

Both the light source and the power meter must remain switched on now until all the measurements have been completed. This is to give the light source a good chance of maintaining a constant light level since the internal temperature and battery voltage will (hopefully) remain stable. The power meter needs to remain on in order to remember the level set as 0 dBr.

Disconnect the patchcord.

Step 2 Connecting them to the system — Figure 15.5

Disconnect the fiber to be tested at the transmitter and plug in the light source. To the far end, connect the power meter. The power meter will immediately show a new figure such as –8.2 dBr. This is the loss over the system. We have actually assumed the patchcord loss (about 0.05 dB) to be small enough to be ignored.

No dBr mode?

Some very basic power meters do not have a dBr facility. Apart from inflicting some calculations on the user, the results will be the same.

Step 1 The power out from the light source

Using such a meter in the above system, we would have read the light source power out in dBm — let's assume this to be –10 dBm.

Step 2 The power out from the link

Leaving the meters switched on, connect them to the optic fiber link. The power meter would now indicate –18.2 dBm.

Step 3 The extra bit

The loss in the optic fiber link is the difference between the two measurements, $(-10.0) - (-18.2) = 8.2$ dB as in the previous case.

Limitations of the light source and power meter method

Assume an optic fiber link has become unreliable. Meters would give an accurate enough result for the overall loss of a system which would confirm that additional losses have occurred. They are, however, unable to indicate which part of the system is responsible for the additional loss. This problem is solved by an *optical time domain reflectometer* (OTDR).

The optical time domain reflectometer (OTDR)

This instrument is connected to one end of any fiber optic system up to 250 km in length. Within a few seconds, we are able to measure the overall loss, or the loss of any part of a system, the overall length of the fiber and the distance between any points of interest. It's really quite amazing.

A use for Rayleigh scatter — Figure 15.6

As light travels along the fiber, a small proportion of it is lost by Rayleigh scattering. As the light is scattered in all directions, some of it just happens to return back along the fiber towards the light source. This returned light is called *backscatter*.

Figure 15.4

Measuring power
from the light
source

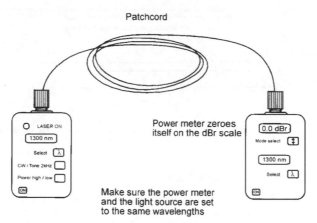

Figure 15.5

Measuring the loss
in the optic fiber
link

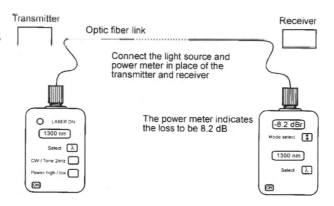

Figure 15.6

Backscatter

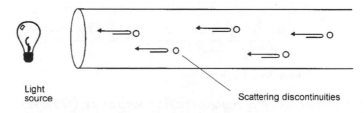

The backscatter power is a fixed proportion of the incoming power and as the losses take their toll on the incoming power, the returned power also diminishes as shown in Figure 15.7.

The OTDR can continuously measure the returned power level and hence deduce the losses encountered on the fiber. Any additional losses such as connectors and fusion splices have the effect of suddenly reducing the transmitted power on the fiber and hence causing a corresponding change in backscatter power. The position and the degree of the losses can ascertained.

Figure 15.7

The backscatter

becomes weaker

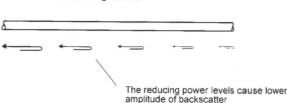

As the light is scattered and absorbed, the power levels decrease as the light travels along the fiber

The reducing power levels cause lower amplitude of backscatter

Measuring distances — Figure 15.8

The OTDR uses a system rather like a radar set. It sends out a pulse of light and 'listens' for echoes from the fiber.

Figure 15.8

How long is the

fiber?

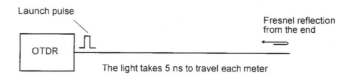

Launch pulse

Fresnel reflection from the end

OTDR

The light takes 5 ns to travel each meter

The reflected light reaches the OTDR 1.4 µs after the pulse was launched

If it knows the speed of the light and can measure the time taken for the light to travel along the fiber, it is an easy job to calculate the length of the fiber.

To find the speed of the light

Assuming the refractive index of the core is 1.5, the infrared light travels at a speed of:

$$\frac{\text{speed of light in free space}}{\text{refractive index of the core}} = \frac{3 \times 10^8}{1.5} = 2 \times 10^8 \text{ ms}^{-1}$$

This means that it will take:

$$\frac{1}{2 \times 10^8} \text{ s or } 5 \text{ ns to travel a distance of 1 meter}$$

This is a useful figure to remember, 5 nanoseconds per meter (5 nsm^{-1}).

If the OTDR measures a time delay of 1.4 µs, then the distance traveled by the light is:

$$\frac{1.4 \times 10^{-6}}{5 \times 10^{-9}} = 280 \text{ m}$$

The 280 meters is the total distance traveled by the light and is the 'there and back' distance. The length of the fiber is therefore only 140 m. This adjustment is performed automatically by the OTDR — it just displays the final result of 140 m.

Inside the OTDR — Figure 15.9

Timer

The timer produces a voltage pulse which is used to start the timing process in the display at the same moment as the laser is activated.

Figure 15.9

A block diagram of an OTDR

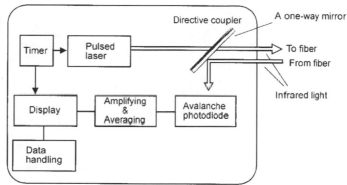

Pulsed laser

The laser is switched on for a brief moment. The 'on' time being between 1 ns and 10 μs. We will look at the significance of the choice of 'on' time or pulsewidth in a moment or two. The wavelength of the laser can be switched to suit the system to be investigated.

Directive coupler

The directive coupler allows the laser light to pass straight through into the fiber under test. The backscatter from the whole length of the fiber approaches the directive coupler from the opposite direction. In this case the mirror surface reflects the light into the avalanche photodiode (an APD). The light has now been converted into an electrical signal.

Amplifying & averaging

The electrical signal from the APD is very weak and requires amplification before it can be displayed. The averaging feature is quite interesting and we will look at it separately towards the end of this chapter.

Display

The amplified signals are passed on to the display. The display is either a cathode ray tube (CRT) like an oscilloscope or a computer monitor, or a liquid crystal as in calculators and laptop computers. They display the returned signals on a simple XY plot with the range across the bottom and the power level in decibels up the side.

Figure 15.10 shows a typical display. The current settings are shown immediately over the grid. They can be modified to suit the measurements being undertaken. The range scale displayed shows a 50 km length of fiber. In this case it is from 0 to 50 km but it could be any other 50 km slice, for example, from 20 km to 70 km. It can also be expanded to give a detailed view of a shorter length of fiber such as 0–5 m, or 25–30 m.

Figure 15.10

An OTDR display –

no signal

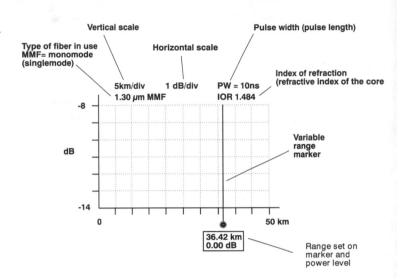

The ranges can be read from the horizontal scale but for more precision, a variable range marker is used. This is a movable line which can be switched on and positioned anywhere on the trace. Its range is shown on the screen together with the power level of the received signal at that point. To find the length of the fiber, the marker is simply positioned at the end of the fiber and the distance is read off the screen. It is usual to provide up to five markers so that several points can be measured simultaneously.

Data handling

An internal memory or a floppy disk drive can store the data for later analysis. The output is also available via an RS232 link for downloading to a computer. In addition, many OTDRs have an onboard printer to provide hard copies of the information on the screen. This provides useful 'before and after' images for fault repair as well as a record of the initial installation.

A simple measurement

If we were to connect a length of fiber, say 300 m, to the OTDR the result would look as shown in Figure 15.11.

Whenever the light passes through a cleaved end of a piece of fiber, a fresnel reflection occurs. This is seen at the far end of the fiber and also at the launch connector. Indeed, it is quite usual to obtain a Fresnel reflection from the end

Figure 15.11

A simple fiber

measurement

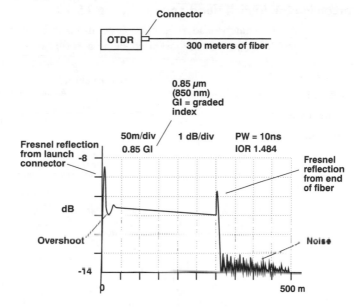

of the fiber without actually cleaving it. Just breaking the fiber is usually enough.

The Fresnel at the launch connector occurs at the front panel of the OTDR and, since the laser power is high at this point, the reflection is also high. The result of this is a relatively high pulse of energy passing through the receiver amplifier. The amplifier output voltage swings above and below the real level, in an effect called ringing. This is a normal amplifier response to a sudden change of input level. The receiver takes a few nanoseconds to recover from this sudden change of signal level.

Dead zones

The Fresnel reflection and the subsequent amplifier recovery time results in a short period during which the amplifier cannot respond to any further input signals. This period of time is called a dead zone. It occurs to some extent whenever a sudden change of signal amplitude occurs. The one at the start of the fiber where the signal is being launched is called the *launch dead zone* and others are called *event dead zones* or just *dead zones*. See Figures 15.12 and 15.20.

Figure 15.12

Dead zone (see

also Figure 15.20)

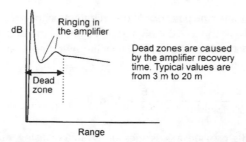

159

Overcoming the launch dead zone — Figure 15.13

As the launch dead zone occupies a distance of up to 20 meters or so, this means that, given the job of checking a 300 m fiber, we may only be able to

Figure 15.13

A patchcord

overcomes dead

zone problems

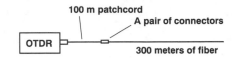

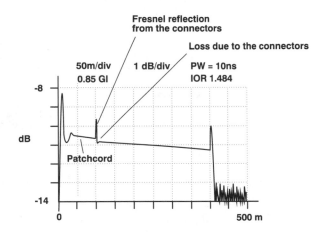

check 280 m of it. The customer would not be delighted.

To overcome this problem, we add our own patchcord at the beginning of the system. If we make this patchcord about 100 m in length, we can guarantee that all launch dead zone problems have finished before the customer's fiber is reached.

The patchcord is joined to the main system by a connector which will show up on the OTDR readout as a small Fresnel reflection and a power loss. The power loss is indicated by the sudden drop in the power level on the OTDR trace.

Length and attenuation

The end of the fiber appears to be at 400 m on the horizontal scale but we must deduct 100 m to account for our patchcord. This gives an actual length of 300 m for the fiber being tested.

Immediately after the patchcord Fresnel reflection the power level shown on the vertical scale is about −10.8 dB and at the end of the 300 m run, the power has fallen to about −11.3 dB. A reduction in power level of 0.5 dB in 300 meters indicates a fiber attenuation of:

$$\frac{\text{attenuation}}{\text{length in kilometers}} = \frac{0.5}{0.3} = 1.66 \text{ dBkm}^{-1}$$

Most OTDRs provide a loss measuring system using two markers. The two markers are switched on and positioned on a length of fiber which does not include any other events like connectors or whatever as shown in Figure 15.14.

Figure 15.14

Using two markers

for loss

measurement

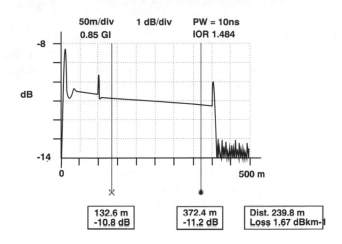

The OTDR then reads the difference in power level at the two positions and the distance between them, performs the above calculation for us and displays the loss per kilometer for the fiber. This provides a more accurate result than trying to read off the decibel and range values from the scales on the display and having to our own calculations.

An OTDR display of a typical system

The OTDR can 'see' Fresnel reflections and losses. With this information, we can deduce the appearance of various events on an OTDR trace as seen in Figure 15.15.

Figure 15.15

A typical OTDR

trace

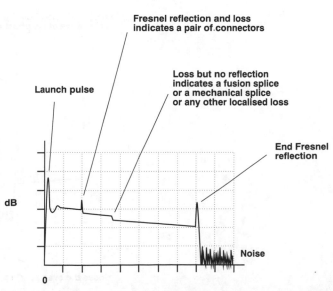

Connectors

A pair of connectors will give rise to a power loss and also a Fresnel reflection due to the polished end of the fiber.

Fusion splice

Fusion splices do not cause any Fresnel reflections as the cleaved ends of the fiber are now fused into a single piece of fiber. They do, however, show a loss of power. A good quality fusion splice will actually be difficult to spot owing to the low losses. Any sign of a Fresnel reflection is a sure sign of a very poor fusion splice.

Mechanical splice

Mechanical splices appear similar to a poor quality fusion splice. The fibers do have cleaved ends of course but the Fresnel reflection is avoided by the use of index matching gel within the splice. The losses to be expected are similar to the least acceptable fusion splices.

Bend loss

This is simply a loss of power in the area of the bend. If the loss is very localized, the result is indistinguishable from a fusion or mechanical splice.

Ghost echoes (false reflections)

In Figure 15.16, some of the launched energy is reflected back from the connectors at the end of the patchcord at a range of 100 m. This light returns and strikes the polished face of the launch fiber on the OTDR front panel. Some of this energy is again reflected to be re-launched along the fiber and will cause another indication from the end of the patchcord, giving a false, or ghost, Fresnel reflection at a range of 200 m and a false 'end' to the fiber at 500 m.

Figure 15.16

Ghosts

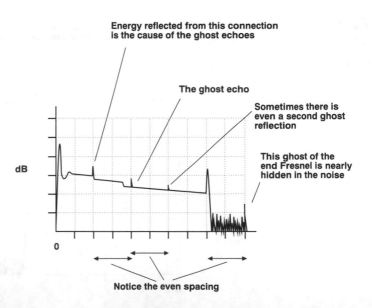

Energy reflected from this connection is the cause of the ghost echoes

The ghost echo

Sometimes there is even a second ghost reflection

This ghost of the end Fresnel is nearly hidden in the noise

dB

0

Notice the even spacing

As there is a polished end at both ends of the patchcord, it is theoretically possible for the light to bounce to and fro along this length of fiber giving rise to a whole series of ghost reflections. In the figure a second reflection is shown at a range of 300 m.

It is very rare for any further reflections to be seen. We have seen earlier that the maximum amplitude of the Fresnel reflection is 4% of the incoming signal, and is usually much less. Looking at the calculations for a moment, even assuming the worst reflection, the returned energy is 4% or 0.04 of the launched energy. The re-launched energy, as a result of another reflection is 4% of the 4% or $0.04^2 = 0.0016$ x input energy. This shows that we need a lot of input energy to cause a ghost reflection. A second ghost would require another two reflections giving rise to a signal of only 0.000 002 56 of the launched energy. Subsequent reflections die out very quickly as we could imagine.

Ghost reflections can be recognized by their even spacing. If we have a reflection at 387 m and another at 774 m then we have either a strange coincidence or a ghost. Ghost reflections have a Fresnel reflection but do not show any loss. The loss signal is actually of too small an energy level to be seen on the display. If a reflection shows up after the end of the fiber, it has got to be a ghost.

Effects of changing the pulsewidth

The maximum range that can be measured is determined by the energy contained within the pulse of laser light. The light has to be able to travel the full length of the fiber, be reflected, and return to the OTDR and still be of larger amplitude than the background noise. Now, the energy contained in the pulse is proportional to the length of the pulse and so to obtain the greatest range the longest possible pulsewidth should be used as illustrated in Figure 15.17.

Figure 15.17

A longer pulse

contains more

energy

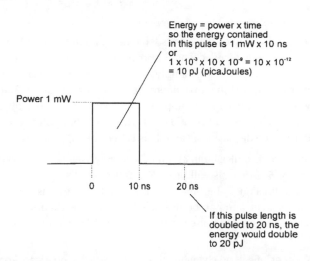

Energy = power x time
so the energy contained
in this pulse is 1 mW x 10 ns
or
$1 \times 10^3 \times 10 \times 10^{-9} = 10 \times 10^{-12}$
= 10 pJ (picaJoules)

Power 1 mW

0   10 ns   20 ns

If this pulse length is doubled to 20 ns, the energy would double to 20 pJ

This cannot be the whole story, as OTDRs offer a wide range of pulsewidths.

We have seen that light covers a distance of 1 meter every 5 nanoseconds so a pulsewidth of 100 ns would extend for a distance of 20 meters along the fiber (Figure 15.18). When the light reaches an event, such as a connector, there is a reflection and a sudden fall in power level. The reflection occurs over the whole of the 20 m of the outgoing pulse. Returning to the OTDR is therefore a 20 m reflection. Each event on the fiber system will also cause a 20 m pulse to be reflected back towards the OTDR.

Figure 15.18

A 100 µs pulse causes a 20 m reflection

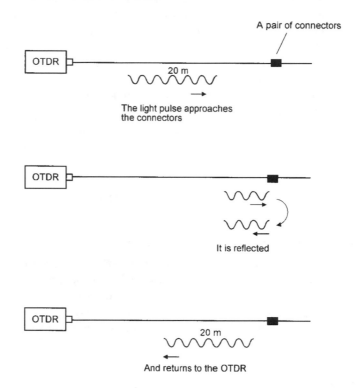

A pair of connectors

The light pulse approaches the connectors

It is reflected

And returns to the OTDR

Now imagine two such events separated by a distance of 10 m or less as in Figure 15.19. The two reflections will overlap and join up on the return path to the OTDR. The OTDR will simply receive a single burst of light and will be quite unable to detect that two different events have occurred. The losses will add, so two fusion splices for example, each with a loss of 0.2 dB will be shown as a single splice with a loss of 0.4 dB. The minimum distance separating two events that can be displayed separately is called the range discrimination of the OTDR.

The shortest pulsewidth on an OTDR may well be in the order of 10 ns so at a rate of 5 nsm^{-1} this will provide a pulse length in the fiber of 2 m. The range discrimination is half this figure so that two events separated by a distance greater than 1 m can be displayed as separate events. At the other end of the scale, a maximum pulsewidth of 10 µs would result in a range discrimination of 1 km.

Another effect of changing the pulsewidth is on dead zones. Increasing the energy in the pulse will cause a larger Fresnel reflection. This, in turn, means

Figure 15.19

Pulse length

determines the

range

discrimination

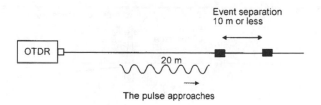

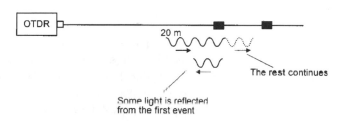

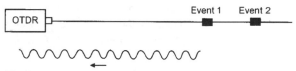

that the amplifier will take longer to recover and hence the event dead zones will become longer as shown in Figure 15.20.

Which pulsewidth to use?

Most OTDRs give a choice of at least five different pulse lengths from which to select.

Low pulsewidths mean good separation of events but the pulse has a low energy content so the maximum range is very poor. A pulsewidth of 10 ns may well provide a maximum range of only a kilometer with a range discrimination of 1 meter.

The wider the pulse, the longer the range but the worse the range discrimination. A 1 μs pulsewidth will have a range of 40 km but cannot separate events closer together than 100 m.

As a general guide, use the shortest pulse that will provide the required range.

Averaging

The instantaneous value of the backscatter returning from the fiber is very weak and contains a high noise level which tends to mask the return signal.

As the noise is random, its amplitude should average out to zero over a period

165

Figure 15.20

Event dead zones

and distances

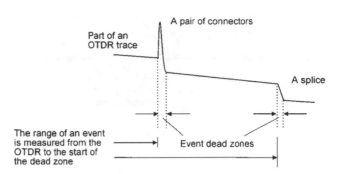

A pair of connectors

Part of an
OTDR trace

A splice

The range of an event
is measured from the
OTDR to the start of
the dead zone

Event dead zones

of time. This is the idea behind the averaging circuit. The incoming signals are stored and averaged before being displayed. The larger the number of signals averaged, the cleaner will be the final result but the slower will be the response to any changes that occur during the test. The mathematical process used to perform the effect is called *least squares averaging* or LSA.

Figure 15.21 demonstrates the enormous benefit of employing averaging to reduce the noise effect. Occasionally it is useful to switch the averaging off to see a *real time* signal from the fiber to see the effects of making adjustments to connectors etc. This is an easy way to optimize connectors, mechanical splices, bends etc. simply fiddle with it and watch the OTDR screen.

Figure 15.21

The benefits of

averaging

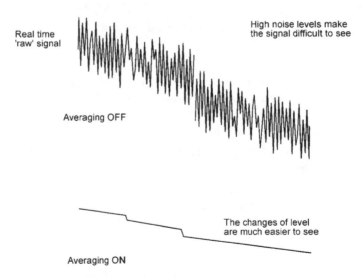

Real time
'raw' signal

High noise levels make
the signal difficult to see

Averaging OFF

The changes of level
are much easier to see

Averaging ON

Dynamic range

When a close range reflection, such as the launch Fresnel occurs, the reflected energy must not be too high otherwise it could damage the OTDR receiving circuit. The power levels decrease as the light travels along the fiber and eventually, the reflections are similar in level to that of the noise and can no longer be used.

The difference between the highest safe value of input power and the minimum detectable power is called the dynamic range of the OTDR and, along with the pulsewidth and the fiber losses, determines the useful range of the equipment.

If an OTDR was quoted as having a dynamic range of 36 dB, it could measure an 18 km run of fiber with a loss of 2 dBkm^{-1}, or alternatively a 72 km length of fiber having a loss of 0.5 dBkm^{-1}, or any other combination that multiplies out to 36 dB.

Fault locator

These devices are used to locate faults quickly and easily rather than provide a detailed analysis of a system and are therefore more likely to be met in a repair environment than at a new installation.

There are two types. The first is similar to the simple light source test.

This device extends the test range of a simple light source up to 5 km using a laser operating at 670 nm which is visible red light. It also provides a 2 kHz test tone which can be picked up by live fiber detectors and provides easy identification of the fiber under test.

If the continuity test fails, it can sometimes show us where the break has occurred. At the position of the break, the light is scattered. Using a powerful visible light source the fiber glows red from the scattered light. This assumes, of course, that the red light is able to penetrate the outer covering of the fiber.

The other device is very similar in principle to the OTDR. It is hand held and operated by internal batteries. It transmits an infrared light pulse along the fiber and 'listens out' for any sudden loss exceeding 0.25 dB, 0.5 dB or whatever value you have selected on the front panel control, all lesser events being ignored.

It doesn't actually provide us with a full OTDR type display but, instead, the result is shown on a liquid crystal display as a simple list of ranges at which losses exceed the selected magnitude occur, together with the total length of the fiber. By comparing this information with the expected values, any discrepancy can be seen and any serious fault is located in seconds.

Quiz time 15

In each case, choose the best option.

1 For contract purposes, test equipment should:

(a) have a valid certificate of calibration

(b) be left switched on at all times

(c) be powered by internal batteries

(d) be manufactured by the same company as produced the fiber

2 Light sources often include a tone output. This enables:

(a) the length of the fiber to be calculated

(b) our eyes to see the infrared light more easily

(c) longer fibers to be tested

(d) easy identification of the fiber being tested

3 For detailed examination of a very short length of optic fiber cable, the OTDR should use the:

(a) shortest possible wavelength

(b) shortest possible pulsewidth

(c) longest available pulsewidth

(d) highest available power

4 Figure 15.22 shows a ghost reflection on an OTDR in option:

Figure 15.22

Question 4

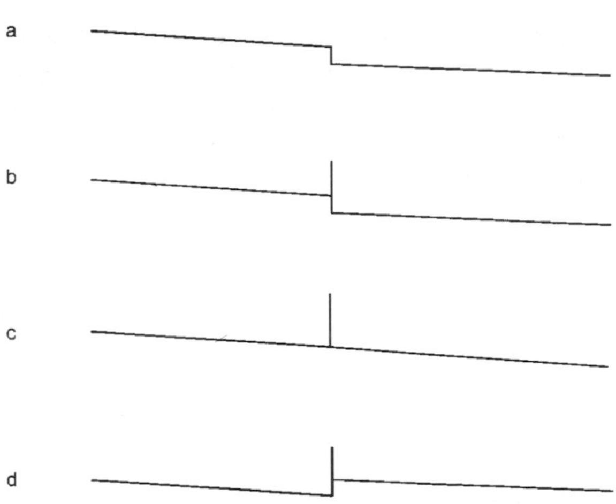

a

b

c

d

(a) a

(b) b

(c) c

(d) d

5   **The averaging facility of an OTDR can be switched off:**

(a) to provide a real time response

(b) to remove noise and clean up the display of information

(c) to allow the peak power to be used

(d) if simultaneous measurements on more than one fiber system are required

16

System design — or, will it work?

In designing a system, there are many factors to take into account. In this chapter we are looking at just two. We aim to satisfy ourselves that the transmitter is powerful enough for the light to reach the far end of the fiber and whether the bandwidth is sufficient to enable the system is carry data at a high enough rate.

Optical power budget or loss budget or flux budget

If the signal is too weak when it reaches the far end of the system the data will be difficult to separate from the background noise. The will cause the number of errors in the received data bits to increase. If an error occurs once in every thousand million bits it would be said to have a *bit error rate* (BER) of 10^{-9} and is the usual lower limit of acceptability.

Limitations on the received power

1 The received power must be high enough to keep the BER to a low value.

2 The received power must be low enough to avoid damage to the receiver.

Limitations on the transmitted power

On cost and safety grounds it is good to keep the transmitter power to the minimum acceptable value.

A problem

Having decided on the receiver and the system, what transmitter power would be needed?

Method of solving

This applies equally well to multimode, singlemode silica fibers and to plastic fibers. The method is summarized in Figure 16.1.

Figure 16.1

How much power

do we need?

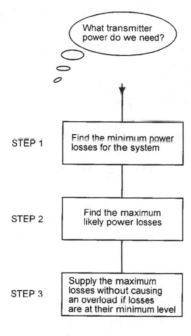

STEP 1 — Find the minimum power losses for the system

STEP 2 — Find the maximum likely power losses

STEP 3 — Supply the maximum losses without causing an overload if losses are at their minimum level

Step 1

Find the minimum power losses due to the:

1. fiber
2. connectors
3. and the splices.

These figures are obtained from the manufacturer either within the catalog or by contacting them.

Step 2

Find the maximum likely losses. This will include the:

1. Minimum losses calculated above.
2. Aging losses. Many components of a system deteriorate during their lifetime and it is important to know how much to allow for this, otherwise the system will crash at some future date. The aging loss is slight in fibers but the transmitter and the connectors, mechanical splices, couplers etc. will need to be

investigated. Again, the manufacturers will supply the data.

3   Repairs. This a matter of judgment depending on the environment and stress to which the fiber will be subjected. It is clearly of little use to design a system that, although it works when first switched on, has so little spare power capacity that the extra loss incurred by a simple repair would be enough to make it fall over. A battlefield system would need more repair allowance that an installation in a building. This stage is really a guess based on experience and advice.

4   Spare. Keep a little extra in reserve — just in case. About 3 dB is a usual amount.

Step 3

Select a transmitter light source with enough power to enable the system to operate under the worse case conditions with the maximum losses considered above. Then check to see if it would damage the receiver in the conditions of minimum loss. If necessary, an attenuator could be added with a view to removing it at a later date should repairs become necessary.

A worked example

As some of the values used are actually guesses (or informed estimates), it is quite to be expected that another person will achieve slightly different results.

Calculate the minimum transmitter power necessary in the system shown in Figure 16.2.

Figure 16.2
What is the
minimum power of
the transmitter?

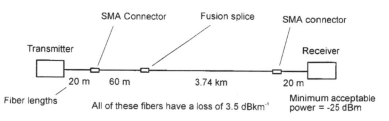

SMA Connector   Fusion splice   SMA connector

Transmitter   Receiver

20 m   60 m   3.74 km   20 m

Fiber lengths   All of these fibers have a loss of 3.5 dBkm⁻¹   Minimum acceptable power = -25 dBm

Step 1 — minimum power loss

The fiber:

The total length of the fiber is:

    20 m + 60 m + 3.74 km + 20m = 3.84 km

The fiber has been checked and found to be of the same specification throughout with a loss of 3.5 dBkm^{-1}.

The total loss is therefore length x loss per unit length. This gives:

    3.84 × 3.5 = 13.44 dB

The connectors:

According to the catalog, the loss for the type used is 1 dB per mated pair.

There are two mated pairs used, so the total loss is:

$2 \times 1 = 2$ dB

The splices:

The fusion splice will be assumed to be 0.2 dB. This is just a guess really.

Total power loss

The minimum value of power loss is therefore just the sum of these losses:

$13.44 + 2 + 0.2 = 15.64$ dB

Step 2

Finding the maximum power loss:

The minimum power loss is 15.64 dB to which the aging and repairs figures must be added.

Aging

Consulting the catalog and where necessary, the suppliers, the aging losses used in this situation are:

- fiber = negligible
- connectors = 0.1 dB per mated pair. Therefore two pairs = 0.2 dB
- splice = negligible
- transmitter = 1 dB
- total aging loss = 1.2 dB

Repairs

A loss of 1.5 dB is assumed. This will vary considerably according to the environment. This is a just a guess.

Spare

We will put in an extra 3 dB loss to take care of unforeseen situations, just in case.

The maximum power loss is:

- minimum loss = 15.64 dB
- aging = 1.2 dB
- repairs = 1.5 dB
- spare = 3 dB

The result is a maximum power loss of:

$15.64 + 1.2 + 1.5 + 3 = 21.34$ dB

Step 3

The transmitter must supply enough power to overcome the worst case losses and still meet minimum power level requirements of the receiver (Figure 16.3).

Figure 16.3

The transmitter output power must be at least −3.66 dBm

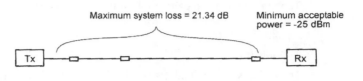

We must be careful to anticipate the likely value of the answer. It is all too easy to get a silly answer by blindly pressing the buttons on the calculator.

The receiver minimum power level is a large negative number of decibels. This means that the power level is very small. The transmitter output power must be greater than this and therefore the numerical value of the decibels must be less negative.

The minimum transmitter power = minimum receiver power + the losses.

So, in our example:

minimum transmitter power = −25 + 21.34 = −3.66 dBm or 430.5 μW

Notice how the magnitude of the losses, 21.34 dB, is added as this represents additional power to be supplied by the transmitter.

The light source with the nearest value of output power available is 500 μW which converts to −3.0 dBm.

Now we need to look at the receiver input under minimum loss conditions as in Figure 16.4.

Figure 16.4

Make sure that the receiver will not be overloaded

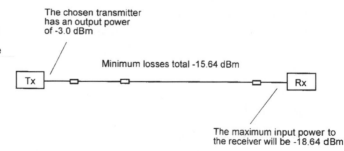

The maximum input power to the receiver occurs when the system losses are at their minimum level.

In our case:

maximum receiver power = transmitter power — minimum losses

So:

maximum receiver power = −3 − 15.64 = −18.64 dBm

In this case the loss figure is reducing the power available and is therefore subtracted.

Hopefully this would not be sufficiently high to damage the receiver. If it would, we could look for a more tolerant one or consider putting an attenuator into the system to use up the excess power. At a later date, due to repairs or modifications to the system, we may need more power. In this case, we could simply remove the attenuator. This also highlights the importance of maintaining adequate records. After several years, no one will remember that an attenuator was fitted into the system.

Another example

We have an existing system. We want to know by how much we could lengthen the cable without having to change the present transmitter and receiver.

Question

Given the system in Figure 16.5, how long could we make the optic fiber cable and still meet the minimum power specification for the receiver.

Figure 16.5

How much cable could we install?

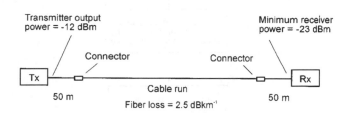

Transmitter output power = -12 dBm

Minimum receiver power = -23 dBm

Connector

Connector

Tx

Rx

50 m

Cable run

Fiber loss = 2.5 dBkm⁻¹

50 m

Method of solving

Step 1

Find the total loss for the known system as it stands at the moment. This is done by gathering the data from catalogs, by measuring, or by consulting the suppliers.

Step 2

Add the aging loss, the likely repair losses and add the spare 3 dB as before.

Step 3

Find the difference in the power of the transmitter output and the minimum receiver power so we know how much power can be used altogether. By subtracting the losses in Steps 1 and 2, we know how much is available for the cable.

Step 4

Once we know the power available for the link, it is an easy job to divide it by the loss per kilometer for the fiber to be used and hence find the safe maximum length for the cable.

A worked solution following the steps itemized in Figure 16.6.

Step 1

The present losses:

Figure 16.6

How long could we make the cable?

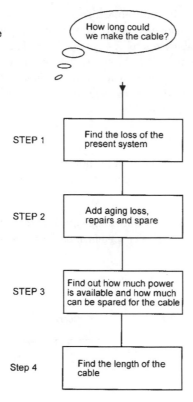

The two patchcords are each 50 m in length giving a total of 100 m. Assuming the fiber used has a quoted loss of 2.5 dBkm^{-1}, this would mean a fiber loss of 0.25 dB.

The mated pairs of connectors are assumed to be 0.2 dB for each pair. Total connector loss is therefore 0.4 dB.

The present losses amount to:

0.25 + 0.4 = 0.65 dB

Step 2

Find the worst case losses for the proposed system.

The aging loss:

0.1 dB per pair of couplers = 0.2 total transmitter 1 dB

For repairs, the assumed value is 2 dB (estimated).

Spare:

3 dB (as usual)

So:

total losses = losses from Step 1 + aging loss + repair power + spare power

In this case:

total losses = 0.65 + 1.2 + 2.0 + 3.0 = 6.85 dB

Step 3

Find the power available for the required cable run.

Transmitter output power is –12 dBm and the minimum receiver input power is –23 dB so there is a total of 11 dB available for the whole system.

Of this 11 dB, we have already accounted for 6.85 dB.

The spare power for the link is:

11 – 6.85 = 4.15 dB

Step 4

The fiber in use has a loss of 2.5 dBkm^{-1}, so the length that could be accommodated is:

$$\frac{4.15}{2.5} = 1.66 \text{ km}$$

As shown in Figure 16.7.

Figure 16.7

The maximum

length of the cable

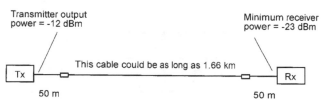

The usable bandwidth of optic fiber

To make a workable system we not only need to be happy with the optical power budget but also be confident that we can achieve a satisfactory data transfer rate and this means having a suitable bandwidth.

If we consult manufacturers' catalogs, bandwidths do not seem too much of a problem. The multimode graded index fiber is quoted as having a bandwidth of 300 MHz to 3 GHz. Single mode is even better at 500 MHz to 10 GHz.

Unfortunately, this tells only half of the story.

We saw in Chapter 7 that dispersion in fiber tends to spread out the pulses of light and they will merge until the information is lost. This puts an upper limit on how fast we can send the data. The magnitude of the dispersion problem increases with the length of the cable and so the transmission rate, and hence the bandwidth, decreases with fiber length.

To make a real system we also need a light source and a receiver. They both have a finite switching speed and limit the transmission rate. There is no point in buying a fiber able to deliver a bandwidth of 10 GHz if the receiver at the far end can only switch at 5 MHz.

Remember it is the system bandwidth that we use, not just the fiber bandwidth so we need to investigate the effect of dispersion and the switching speed of the transmitter and the receiver.

Multimode and singlemode fibers respond a little differently and they need to be considered separately. The multimode case is slightly easier so we will look at that first.

The bandwidth of a multimode system

The method is summarized by Figure 16.8.

Step 1

Find the bandwidth of the fiber

Figure 16.8

Finding the real bandwidth in a multimode system

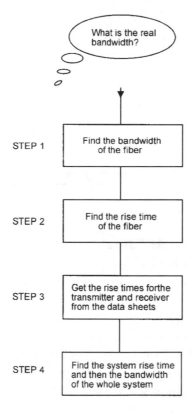

We do this by obtaining the fiber bandwidth from its specification from the data sheet or catalog. A typical specification would be 500 MHz km (sometimes written as 500 MHz.km).

Note the units used 500 MHz km, this means that the bandwidth multiplied by the distance in kilometers comes to 500 MHz. So this particular fiber has a bandwidth of 250 MHz if it were only 2 km in length, but reduces to only 50 MHz if it were 10 km long. Notice already how the dream of a 500 MHz system is fading.

Step 2

Find the fiber rise time (check Figure 14.3 for a reminder on rise time if necessary)

The rise time is the time taken for the light to increase from 10% to 90% of its final value. The higher the frequency of a waveform the shorter will be the rise time as shown in Figure 16.9. It is linked to the bandwidth as this is a measure of the highest frequency component in the signal transmitted along the fiber.

Figure 16.9

Short rise times need wide bandwidths

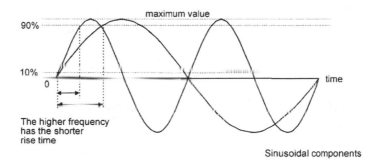

maximum value

90%

10%

0

time

The higher frequency has the shorter rise time

Sinusoidal components

The rise time of the fiber is calculated using the bandwidth of the fiber that we found in Step 1:

$$t_r = \frac{0.35}{\text{bandwidth of the fiber}}$$

where t_r is an abbreviation for rise time. This formula, in common with some other system formulas, has been introduced without proof to avoid the usual long tedious justifications.

Step 3

We also need to know how fast the transmitter and the receiver are able to respond. Obviously, a sluggish device could spoil everything.

We find the rise time in the manufacturers' data. It is shown as response time or as t_r for rise time. A typical value for a laser is 0.3 ns and for an LED, 5 ns. Notice how much faster the laser is. Some receivers particularly phototransistors are not very brisk at all and need to be selected with care.

Step 4

We now have three different components, the light source, the fiber and the receiver, each with their own switching speed. When the electronic pulse is applied to the light source it will start to increase its light output power. The fiber follows at its own rate and finally the receiver electrical output starts to rise. It is not easy to see how these various reaction speeds would interact to produce an overall response for the system.

To do this, we combine the separate rise times:

179

$$t_{r_{sys}} = \sqrt{t_{r_{RX}}^2 + t_{r_{TX}}^2 + t_{r_{FIBER}}^2}$$

This gives the rise time for the overall system.

Now we can get the bandwidth. To do this, we use the same basic formula as we used in Step 2, simply transposed to give the system bandwidth:

$$\text{Bandwidth} = \frac{0.35}{t_{r_{SYS}}}$$

A worked example — multimode

Find the usable bandwidth in the fiber optic system shown in Figure 16.10.

Step 1

Find the bandwidth of the fiber:

Figure 16.10

What is the usable bandwidth of this system?

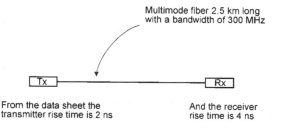

Multimode fiber 2.5 km long with a bandwidth of 300 MHz

From the data sheet the transmitter rise time is 2 ns

And the receiver rise time is 4 ns

$$\text{Fiber bandwidth} = \frac{\text{quoted bandwidth}}{\text{length in km}}$$

$$\text{Fiber bandwidth} = \frac{300 \times 10^6}{2.5} = 120\,\text{MHz}$$

Step 2

Find the fiber rise time:

$$\text{Rise time} = t_{r_{FIBER}} = \frac{0.35}{\text{fiber bandwidth}}$$

$$t_{r_{FIBER}} = \frac{0.35}{120 \times 10^6} = 2.9\,\text{ns}$$

Step 3

Consult the data sheets to find the rise time of the light source and the receiver.

We will assume realistic values of 2 ns for the transmitter rise time and 4 ns for the receiver.

Step 4

We now find the system rise time by combining the three previous values.

To keep the calculation as simple as possible, we can work in nanoseconds throughout and simply remember to call the answer nanoseconds:

$$t_{r_{sys}} = \sqrt{t_{r_{RX}}^2 + t_{r_{TX}}^2 + t_{r_{FIBER}}^2}$$

$$t_{r_{sys}} = \sqrt{4^2 + 2^2 + 2.9^2} \quad (\text{all values in ns})$$

$$t_{r_{sys}} = 5.3 \text{ ns}$$

Now we can find the system bandwidth:

$$\text{System bandwidth} = \frac{0.35}{t_{r_{SYS}}}$$

$$\text{System bandwidth} = \frac{0.35}{5.3 \times 10^{-9}} = 66 \text{ MHz}$$

as seen in Figure 16.11.

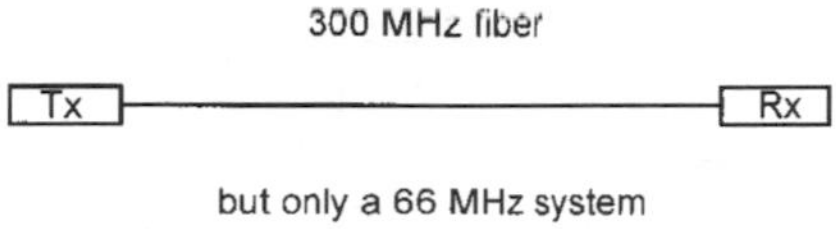

Figure 16.11

The useful bandwidth doesn't only depend on the fiber

300 MHz fiber

Tx —————————————— Rx

but only a 66 MHz system

It's interesting to see how the fiber is only one factor that determines the performance of the system. Our 300 MHz optic fiber has actually produced a system with a bandwidth of only 66 MHz. What a disappointment!

The bandwidth of a singlemode system

We mentioned earlier that there are a couple of differences when it came to finding the bandwidth of a singlemode system.

When we look at data sheets, we see that there are no bandwidth figures quoted for singlemode fiber. Instead, they provide a figure for the dispersion characteristics and from this, we must calculate the bandwidth. It is not possible for the manufacturers to do this for us because the dispersion is dependent on the spectral width of the light source and the length of the fiber, and both of these are beyond their control.

Compared with the multimode calculation, we replace Step 1 but thereafter everything else is the same. The process is summarized in Figure 16.12.

Step 1

Find the dispersion

The dispersion depends on the dispersion specification of the fiber, the spectral width of the light source and the fiber length.

To find the dispersion we use the formula:

dispersion = dispersion specification for the fiber × spectral width of the light source × length of the fiber

181

Figure 16.12

A singlemode
system — the real
bandwidth

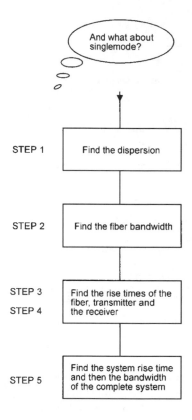

STEP 1 — Find the dispersion

STEP 2 — Find the fiber bandwidth

STEP 3 / STEP 4 — Find the rise times of the fiber, transmitter and the receiver

STEP 5 — Find the system rise time and then the bandwidth of the complete system

The combination of these three factors, dispersion in picoseconds, spectral width in nanometers and length in kilometers produces the impressive unit of $psnm^{-1}km^{-1}$ (read as picoseconds per nanometer per kilometer).

Step 2

Find the fiber bandwidth

This is simple enough:

$$\text{Fiber bandwidth} = \frac{0.44}{\text{dispersion figure}}$$

The resultant figure is usually in GHz but, just as in multimode fiber, disappointment often follows unless the transmitter and receiver are carefully chosen.

The remaining stages of the calculation follows exactly the same pattern as we used for multimode fiber but it is outlined here for convenience.

Step 3

Find the fiber rise time.

The rise time of the fiber is calculated using the bandwidth of the fiber that we found in Step 2:

$$t_r = \frac{0.35}{\text{bandwidth of the fiber}}$$

Step 4

We find the rise time in the manufacturer's data. A typical value for a laser is 0.3 ns and for an LED is 5 ns. This is another advantage of using a laser for a singlemode system where data rates are likely to be of importance.

Step 5

We now move on to find the overall response for the system.

As before, we combine the separate rise times:

$$t_{r_{sys}} = \sqrt{t_{r_{RX}}^2 + t_{r_{TX}}^2 + t_{r_{FIBER}}^2}$$

This gives us the rise time for the overall system.

Now we can get the bandwidth. To do this, we use the same basic formula as we used in Step 3, simply transposed to give the system bandwidth:

$$\text{Bandwidth} = \frac{0.35}{t_{r_{SYS}}}$$

A worked example

Calculate the system bandwidth for the singlemode system shown in Figure 16.13.

Figure 16.13

The information needed to find the bandwidth of a singlemode system

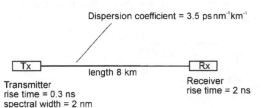

Dispersion coefficient = 3.5 ps nm^{-1}km^{-1}

Tx — length 8 km — Rx

Transmitter
rise time = 0.3 ns
spectral width = 2 nm

Receiver
rise time = 2 ns

Step 1

Find the dispersion

To find the dispersion we use the formula:

dispersion = dispersion specification for the fiber × spectral width of the light source × length of the fiber

Enter the values in pica seconds, nanometers and kilometers.

dispersion = 3.5 × 2 × 8 = 56 ps nm^{-1} km^{-1}

Step 2

Find the fiber bandwidth

$$\text{Fiber bandwidth} \quad = \quad \frac{0.44}{\text{dispersion figure}}$$

$$\text{Fiber bandwidth} \quad = \quad \frac{0.44}{56 \times 10^{-12}}$$

$$\text{Fiber bandwidth} \quad = \quad 7.86 \text{ GHz}$$

Step 3

Find the fiber rise time

$$t_r = \frac{0.35}{\text{bandwidth of the fiber}}$$

$$t_r = \frac{0.35}{7.86 \times 10^9} = 44.53 \text{ ps}$$

Step 4

From the data supplied, the transmitter rise time is 0.3 ns in this example, and the receiver rise time is 2 ns.

Step 5

The overall response for the system is given by the formula:

$$t_{r_{sys}} = \sqrt{t_{r_{RX}}^2 + t_{r_{TX}}^2 + t_{r_{FIBER}}^2}$$

We can now enter our data but remember that the rise time of the fiber is in picoseconds and not nanoseconds. In order to be able to work in nanoseconds, we need to change the picoseconds into nanoseconds by dividing it by 1000 to give 0.0445 ns:

$$t_{r_{sys}} = \sqrt{2^2 + 0.3^2 + 0.0445^2}$$

$$t_{r_{sys}} = \sqrt{4.09}$$

$$t_{r_{sys}} = 2.02 \text{ ns}$$

This gives us the rise time for the overall system.

Now we can find the system bandwidth:

$$\text{System bandwidth} \quad = \quad \frac{0.35}{t_{r_{SYS}}}$$

$$\text{System bandwidth} \quad = \quad \frac{0.35}{2.02 \times 10^{-9}} = 173.3 \text{ MHz}$$

The lesson to be learned here is that the light source performance and the receiver rise time are really significant if the full potential of a singlemode system is to be achieved.

In this example system, by looking at the system rise time formula:

$$t_{r_{sys}} = \sqrt{2^2 + 0.3^2 + 0.0445^2}$$

we can see that the 2 ns due to the receiver response time is significantly greater than the 0.3 ns of the laser and the very small 0.0445 due to the fiber. It follows then, that little would be achieved by spending more money on better fiber or indeed a faster laser. Changing the receiver to one with a 1 ns response time would increase the system bandwidth to approximately 335 MHz. A good investment.

Quiz time 16

In each case, choose the best option.

1 Reducing the length of a multimode fiber would:

(a) decrease the transmitter rise time

(b) decrease the system bandwidth

(c) increase the bandwidth of the system

(d) increase the system rise time

2 Changing the spectral width of the light source would affect the:

(a) fiber bandwidth in a singlemode system

(b) system bandwidth of a multimode system but not a singlemode one

(c) aging losses

(d) number of likely repairs

3 If the transmitter and the receiver rise times were 0.5 ns and 1.5 ns respectively, and the fiber rise time was 25 ps, the system rise time would be approximately:

(a) 25.05 ns

(b) 1.42 ns

(c) 1.58 ns

(d) 5.19 ns

4 Referring to the multimode system in Figure 16.14, the maximum usable length of fiber would be approximately:

(a) 1.2 km

(b) 2.2 km

(c) 3.2 km

(d) 4.2 km

Figure 16.14

Question 4

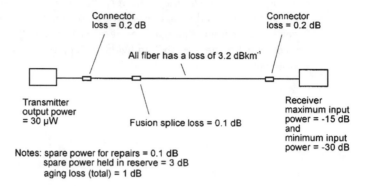

Connector loss = 0.2 dB

Connector loss = 0.2 dB

All fiber has a loss of 3.2 dBkm^{-1}

Transmitter output power = 30 μW

Fusion splice loss = 0.1 dB

Receiver maximum input power = -15 dB and minimum input power = -30 dB

Notes: spare power for repairs = 0.1 dB
   spare power held in reserve = 3 dB
   aging loss (total) = 1 dB

5   If the dispersion figure for a singlemode fiber is known to be 44 ps, the bandwidth of the fiber would be:

(a) 10 GHz

(b) 7.95 GHz

(c) 100 MHz

(d) 795 MHz

17

The transmission of signals

So far we have been concerned with the design and operation of fiber optic systems for the transmission of light. For it to be of use to us, we must be able to send data along the fiber and be able to recover it at the far end.

We will now have a look at some alternative techniques.

Analog transmission

This is the simplest method and is generally used for short range work. It is illustrated in Figure 17.1.

Figure 17.1

An analog

transmission system

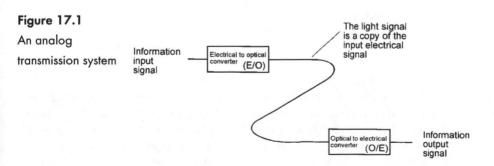

The incoming information signal, speech, music, video etc. is used to control the power output from the LED or the laser. The light output is, as near as possible, a true copy of the electrical variations at the input. At the far end of the fiber, the receiver converts the light back into an electrical signal which is, hopefully, the same as the original electrical signal.

Any nonlinearity of the characteristics of the transmitter or receiver will reduce the accuracy of the electrical/optical (E/O) and optical/electrical (O/E) conversions and give rise to distortion in the output signal.

Another problem is noise. Since the receiver is receiving an analog signal, it must be sensitive to any changes in amplitude. Any random fluctuations in light level caused by the light source, the fiber or the receiver will cause unwanted noise in the output signal. Electrical noise due to lightning will also give rise to electrical noise in the non-fiber parts of the system.

As the signal travels along the fiber, it is attenuated. To restore signal amplitude, we add amplifiers called repeaters at regular intervals. The repeater has a limited ability to reduce noise and distortion present which means that we tend to accumulate problems as we go.

Digital transmission

In a digital system, the information signal is represented by a sequence of on/off levels. We will see how this is done in a minute or two. The 'on' state is often referred to as logic 1 and the 'off' state as logic 0. The 1 and 0 have no numerical significance and are just convenient ways to differentiate between the two states. We could have called them black and white, up and down, salt and pepper, or anything else.

At the repeaters, and at the final receiver we only have to ask whether or not a signal is present by comparing the signal with a chosen *threshold* level shown in Figure 17.2.

Figure 17.2

Repairing damaged digital pulses

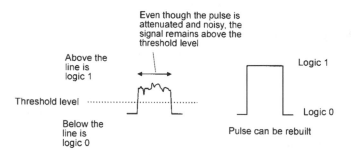

This 'yes' or 'no' approach means that it ignores noise and distortion since all voltages above the threshold level are recognized as a logic 1 state and all below this level as a logic 0. The signal is then regenerated as a perfect copy of the original signal. Repeaters in digital systems are called regenerators since they regenerate or rebuild the signals rather than just amplifying them. Regenerator spacing can be increased allowing signals to become very small before there is any danger of it becoming lost in the noise.

Figure 17.3 offers a summary of analog and digital systems.

Figure 17.3

One of the benefits

of a digital system

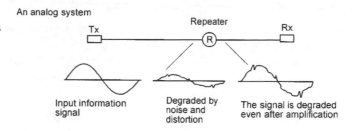

An analog system

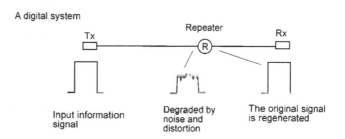

A digital system

Converting an information signal into a digital form (digitizing)

Sampling — Figure 17.4

The first step in the process is to measure the amplitude of the input waveform at regular intervals. These measurements are called *samples*.

It has been found that providing the samples are taken at frequent enough intervals they will contain sufficient information about the input signal for it to be reformed at the far end of the system.

The sampling rate must be greater than twice the highest frequency contained in the input signal. This is called the *Nyquist criterion*. For example, a typical telephone link would contain frequency components up to a maximum of 3.4 kHz and could be sampled at any frequency greater than 6.8 kHz.

Pulse amplitude modulation (PAM)

The transmission is a series of pulses and at first glance it appears to be digital but it is actually an analog signal in disguise. The deciding feature that makes it analog is that the sample pulses can be of any amplitude whereas a digital signal is a series of pulses all of which are of the same size.

PAM is rarely used since we have had to go to all the trouble of sampling the signal but we still have an analog signal and is therefore still susceptible to noise. It seems like we have gone to a lot of trouble for nothing. However it is the essential first stage in converting the signal into a digital format.

Pulse code modulation (PCM)

In a PCM system, the information signal is sampled as in the PAM system, and the amplitude of each sample is then represented by a binary code. In Figure

Figure 17.4

Pulse amplitude

modulation (PAM)

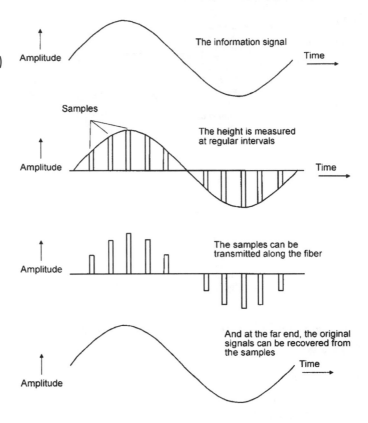

17.5 the maximum amplitude is divided into 8 levels counting up from 0 to 7 in binary. In some cases, the value of the waveform will fall between two values such as in sample 3 where it is between 110 and 111. Whether the coding circuit decides to value it as 110 or as 111 it will not be a true representation of the information signal. This misrepresentation is called *quantization noise*.

Quantization noise can be reduced by increasing the number of levels used but never eliminated. For example, if an 8 volt signal were to be quantized into 8 levels, the levels would increase in 1 volt steps. If the nearest level was always selected, the quantization noise could have a maximum value of 0.5 volt being half the step value. Now, if we increased the number of steps to say 80, the levels would be 0.1 volt apart and the maximum possible error would again be half of the step value, or 0.05 volts. Increasing the number of steps means less quantization noise at the expense of having to send more binary digits and hence increasing the transmission bandwidth.

Within the time interval associated with each sample, called a time slot, the binary code is transmitted as a digital code. In Figure 17.5, the three samples would be transmitted as 011, 010, and 111 (possibly 110). The resulting transmission is shown in Figure 17.6.

At the receiver, the incoming waveform is tested during the sample time to see if the signal is at logic 0 or at logic 1. The data associated with each sample and

Figure 17.5

PCM coding

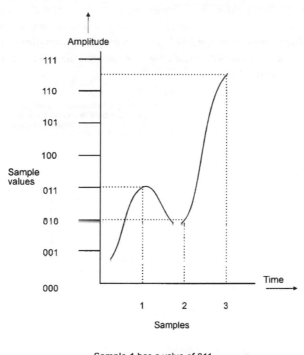

Sample 1 has a value of 011
Sample 2 has a value of 010

But what about sample 3?

Figure 17.6

PCM transmission

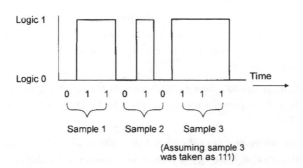

Sample 1   Sample 2   Sample 3

(Assuming sample 3
was taken as 111)

hence the original information can then be recovered. This relies on the sampling times being synchronized at the transmitter and at the receiver. There are many different ways to achieve this. The least inviting is to send another signal just to provide timing or *clock* information. This is clearly a wasteful method and, in practice, the clock information is extracted from the incoming signal.

Generally, the level changes that occur in a PCM transmission are used to keep the receiver clock synchronized to the transmitter. Every time the transmitted signal changes its level, the sudden change in voltage is detected by the receiving circuit and used to ensure that the receiver clock remains locked on to the transmitter clock. There are problems when the incoming signal stays at a low level or a high level for a long period of time as shown in Figure 17.7. The constant voltage level gives an output of continuous ones or zeros and the timing information would be lost and the receiver clock would drift out of synchronization.

Figure 17.7

Constant level

signals can cause

timing problems

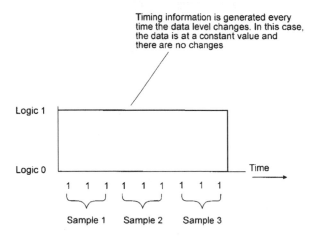

Timing information is generated every time the data level changes. In this case, the data is at a constant value and there are no changes

Logic 1

Logic 0

Time

1 1 1   1 1 1   1 1 1

Sample 1   Sample 2   Sample 3

Biphase (Manchester) code

There are several different methods of coding the data prior to a PCM transmission that overcomes the problem of continuous levels. The Biphase (Manchester) code is a popular one and serves to illustrate the possibilities.

The encoding method is to reverse the level of each pulse during the second half of each timeslot. For example to transmit a 0 level, we send a 0 for half of the available time, and then send a 1. Likewise, to send a level 1, we send a 1 followed by a 0.

Whatever the data being sent there will always be a change of level during each time slot to provide the necessary clock synchronization information (Figure 17.8).

At the receiver a circuit called a decoder is used to reconstruct the PAM signal and from that the original information can be extracted.

How to get more digital signals on a single fiber

Time division multiplexing (TDM) – Figure 17.9

This is the most popular method for increasing the number of signals that can be carried on a single fiber.

If the width of the pulses in a PCM system are reduced, then there will be spaces between them. These spaces can be filled with another transmission containing data associated with another, quite separate, information signal.

The further we reduce the width of the transmitted pulses, the larger the number of other signals that can be slotted in to share the same fiber.

Figure 17.8

Biphase
(Manchester)
coding

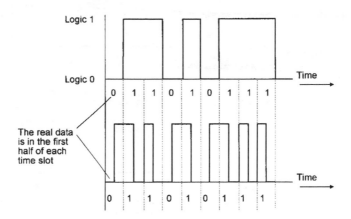

Logic 1

Logic 0

Time

0 1 1 0 1 0 1 1 1

The real data
is in the first
half of each
time slot

Time

0 1 1 0 1 0 1 1 1

Plenty of level changes to
provide timing information

Figure 17.9

Time division
multiplexing (TDM)

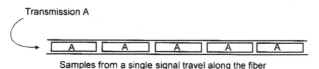

Transmission A

A A A A A

Samples from a single signal travel along the fiber

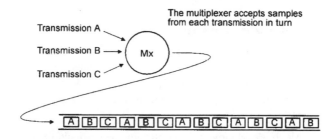

Transmission A

Transmission B → Mx

Transmission C

The multiplexer accepts samples
from each transmission in turn

A B C A B C A B C A B C A B

Using shorter samples, several signals can be sent on the same fiber

The interleaving of the samples from the two transmissions is carried out by a circuit called a *multiplexer*.

At the receiving end we use a *demultiplexer* which simply sorts the samples out into their separate signals from which the information can be recovered.

As the pulses become narrower, the bandwidth of the signal increases and this serves to put an upper limit on the number of signals that can be multiplexed by this method.

Wave division multiplexing (WDM) — Figure 17.10

The technique allows several different signals to be carried along a single fiber at the same time or a fiber to be used for two-directional transmission. It achieves this by using different wavelengths for each transmission and can be

Figure 17.10

Two benefits of wavelength division multiplexing (WDM)

Send on wavelength 1

Receive wavelength 1

Receive wavelength 2

Send on wavelength 2

Transmissions can go in both directions at the same time

Light at several different wavelengths

The transmissions are separated at the receiver

Many transmissions can travel along a single fiber

employed on singlemode or multimode fibers using lasers or LEDs.

We can derive the different frequencies from independent light sources or we can take the light from a broadband device like an LED and split the light into separate channels. The combination and separation can be achieved by wavelength sensitive couplers.

Advocates of this system suggest an increased capacity of between ten and a hundred fold. It is also a simple way to expand an existing fiber link, just send another signal at a different wavelength along the same route.

How to send data further

To overcome the effects of attenuation in the fiber, we must introduce repeaters at intervals to amplify the signals. This involves a light detector to convert the signal from light to an electrical signal. The electrical signal is then amplified and used to power another light source for the next section of fiber.

Much development is continuing on an alternative approach. Optic fiber amplifiers enable a signal to be amplified without converting it to an electrical signal first, a direct light in-light out repeater.

The heart of the matter is a length of very special fluoride based fiber. It is constructed in the same manner as the standard silica fiber but it has the property of being able to transfer light energy between two different signals which are traveling in the same direction in the fiber (Figure 17.11).

Figure 17.11

An optic fiber amplifier

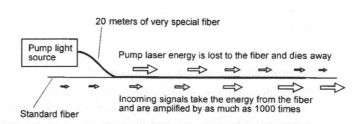

20 meters of very special fiber

Pump light source

Pump laser energy is lost to the fiber and dies away

Standard fiber

Incoming signals take the energy from the fiber and are amplified by as much as 1000 times

We use a light source called a pump to inject energy into the fluoride fiber. The pump light and the weak incoming signal pass through the fiber and as it does so, energy is passed from the pump light into the molecules of the fiber and from there into the signal to be amplified. The wavelength of the pump is critical and is determined by the design of the amplifying fiber in use.

As the light energy travels along the fiber, the pump light becomes weaker as the energy is absorbed by the fiber and the incoming signal is amplified by absorbing energy from the fiber.

A length of about 20 meters can provide an amplification of up to +30 dB (x1000) giving the possibility of using repeaters at distances of well over 100 km.

Quiz time 17

In each case, choose the best option.

1   PAM is:

(a) a pulse system immune from noise

(b) a digital system

(c) a popular system for telephone transmissions

(d) an analog system

2   A suitable sampling rate for a signal containing frequency components between 40 Hz and 4 kHz would be:

(a) 10 kHz

(b) 80 Hz

(c) 90 Hz

(d) 4.04 kHz

3   Quantization noise can be reduced by:

(a) taken samples at a faster rate

(b) using a PCM system

(c) increasing the number of levels during sampling

(d) using higher amplitude signals

4   A bi-phase (Manchester) system:

(a) is not a real digital system

(b) generates timing information, even when the incoming data is held at a constant level

(c) eliminates quantization noise

(d) does not require a receiver clock

5   TDM:

(a) requires a light source able to produce light at more than one wavelength

(b) allows more that one signal to be transmitted along a single optic fiber

(c) cannot be used to transmit digital signals

(d) stands for time duplication multiplexing

18

Some final thoughts

Sitting in the stream

One problem with learning any technology-based subject is that we can never say the job's done. We close our eyes for a moment then look up to find that technology has galloped off into the distance.

What is the best way to learn a language? The answer is obvious — go and live in a country where they speak it and be immersed in it. This is called *sitting in the stream* and applies equally well to any subject and is by far the easiest way of learning about fiber optics, or indeed, anything else.

We simply have to track down sources of anything about our chosen subject and let it trickle past in a gentle stream of information. With no effort at all, some of it will 'stick' whether we apply ourselves or not.

Generating a stream

Exhibitions

Go to any fiber optic exhibitions available. Wander around preferably with a companion, looking at anything of interest and picking up leaflets and catalogs whenever they are available. Attend any demonstrations or seminars that are available. Don't worry about all the words that mean nothing at all — they don't matter. We are not sitting an exam. We are just splashing around in a stream. Bring all the information home and read every word of it, even the adverts, they are an important source of information and are guaranteed to be up to date.

Catalogs

Catalogs are invaluable. They contain lots of information about the products and their specifications. They should be dipped into whenever a few seconds are going spare.

Magazines

Read all magazines that have any articles on fiber optics. Many electronics magazines have the odd article of interest.

There is a magazine called *F.O.C.U.S.* which is devoted entirely to fiber optics and associated topics. An excellent source of information on developments, devices and 'what's going on' generally.

Contacts are:

North America and Canada: Willy R Mattes, 8 Scenic Millway, Toronto M2L 1S3 Canada.

Australia: F.O.C.U.S. subscriptions, Light Networking, PO Box 14, Belgrave, Vic 3160, Australia.

UK: F.O.C.U.S. Ltd., Cotswold House, Kingston, Ringwood Hants BH24 3BQ UK.

India: F.O.C.U.S. subscriptions, Essen Deinki, 22 Industrial Area Phase 2, Chandigarh, India.

South Africa: Chevron Consultants, PO Box 98466, Sloane Park, 2152 Johannesburg, Republic of South Africa.

Training courses

Short courses are available from colleges, universities and industrial training companies. These courses are invaluable since they provide hands-on experience.

Read the course information carefully to ensure that it will do what you want. What is the balance between theory and practice? At what level is the theory pitched?

Bearing in mind that it is the practical work that will probably be the main attraction, we need to ask some questions such as:

How many people are going to attend the course and how many pieces of equipment, e.g. fusion splicers are there available? It's no good going on a course and finding that most of the time is spent waiting around to use the equipment. Demonstrations are useful but are no substitute for actually making our own mistakes.

A reference book

Optical Fiber Communications by John M. Senior. Published by Prentice Hall (1992) ISBN number 0-13-635426-2.

This book provides a detailed theoretical treatment of the subject. It is a standard reference work on the subject. Over 900 pages of solid information. Worth keeping within reach but it is not light holiday reading.

Glossary

Absorption loss

Loss of light in a fiber due to impurities.

Acceptance angle

The largest angle of incident light that lies within the cone of acceptance.

Amplification

An increase in power level measured at two points. Usually measured in the decibels.

Analog

A data format which allows smooth changes of amplitude using all intermediate values.

Angle of incidence

The angle between the incident ray and the normal.

Attenuation

A decrease in power level between two points. Usually expressed in decibels. Opposite to amplification.

Avalanche diode

A device to convert light into an electrical current.

Axial ray

The ray that passes straight through the center of a fiber without being refracted.

Backscatter

The small proportion of light scattered by Rayleigh scattering which is returned towards the source.

Bandwidth

The range of modulation frequencies that can be transmitted on a system while maintaining an output power of at least half of the maximum response.

Bend loss

Losses due to bends in the fiber. The loss increases as the bend radius decreases. See *macrobend* and *microbend*.

Bend radius

Minimum bend radius. The smallest acceptable bend for a fiber or cable before bend loss is apparent.

BER

Bit error rate. The proportion of incoming bits of data that are received incorrectly.

Bit

Abbreviated version of binary digit.

Buffer

See *primary coating*.

Cable

One or more optic fibers contained in a jacket usually also containing strength members etc.

Chromatic dispersion

Dispersion caused by different wavelengths contained in the transmitted light.

Cladding

The clear material surrounding the core of an optic fiber. It has a lower refractive index than the core.

Coherent bundle

A group of optic fibers in which each fiber maintains its position relative to the other fibers so that images can be transmitted along the bundle.

Cone of acceptance

The cone formed by the angles of the light able to enter the fiber core measured from the center line of the core.

Connector

A means of joining optic fibers in a way that allows easy disconnection. In conjunction with an adapter, it performs the same function as a plug and socket in copper based systems.

Core

The central part of the fiber through which most of the light is transmitted. It has a higher refractive index than the cladding.

Coupler

A device to combine several incoming signals onto a single fiber or to split a single signal onto several fibers in a predetermined power ratio.

Critical angle

The lowest angle of the light ray measured with respect to the normal that can be reflected by a change in refractive index.

dB

Abbreviation for decibel. A logarithmic unit used to compare two power levels.

dBm

A power level compared with 1 milliwatt.

dBr

A power level compared with another, stated power level

Decibel

A logarithmic unit used to compare two power levels.

Digital

A data format in which the amplitude can only change by discrete steps.

Dispersion

The widening of light pulses on an optic fiber due to different propagation velocities of the pulse components.

Dolly

A polishing guide used to support a connector during the polishing process.

Dynamic range

The range of usable power levels expressed in decibels.

Eccentricity

Core eccentricity. The amount by which core is not placed centrally within the cladding.

Ferrule

A rigid tube used to confine and support the stripped end of a fiber as found in connectors.

Fiber

An abbreviation for optic fiber or fiber optic.

Fiber optic

Fiber optic system. A communication system using optic fibers. Often abbreviated to *fiber*.

Fresnel reflection

A reflection that occurs from a surface whenever there is a sudden change in the refractive index as at the end of a fiber. The *s* is not pronounced.

Fusion splice

A low loss, permanent means of connecting two fibers involving heating the fibers until they fuse together.

Graded index fiber

A fiber in which the refractive index of the core is at a maximum value at the center and decreases towards the cladding.

Index matching gel

Index matching fluid. A material with a refractive index close to optic fibers and used to reduce the amplitude of Fresnel reflections in couplers, mechanical splices etc.

Insertion loss

The loss of power due to the insertion of a device.

Laser

A light source of low spectral width.

LED

Light emitting diode. A semiconductor device used as a low power light source. The spectral width is greater than a laser.

Loose tube construction

A cable in which the optic fibers are contained loosely in a tube.

Macrobend

A bend in a fiber with a radius of curvature less than the recommended value. It causes a localized power loss which can be eliminated by straightening the fiber.

Mechanical splice

A method of connecting fibers usually involving adhesive and mechanical support and alignment of the fibers.

Meridional ray

A ray which always passes through the core axis as it is propagated.

Micro

A prefix indicating a millionth.

Microbend

A tight bend or kink in the core over distances of a millimeter or less giving rise to a loss.

Micron

A unit of distance, one millionth of a meter, the preferred unit is the micrometer.

milli

A prefix indicating one thousandth.

Modes

Separate optical waves capable of being transmitted along a fiber. The number of modes with a given light wavelength is determined by the NA and the core diameter.

Monomode fiber

Alternative name for singlemode fiber.

Multimode fiber

An optic fiber able to propagate more than one mode at the same time.

Multiplexing

The transmission of several different signals along a single fiber.

nanometer

One thousandth of a micrometer.

nm

Abbreviation for nanometer.

Normal

A line drawn at right angles to the position of a change in refractive index e.g. between the core and the cladding.

Numerical aperture (NA)

The sine of the critical angle between the core and the cladding.

Optic fiber

The length of clear material that can be used to transmit light. Often abbreviated to *fiber*.

Patchcord

A short cable, usually terminated by a connector at each end used to reconfigure a route as required.

Photodiode

A semiconductor device that converts light into an electrical current.

pico

a prefix indicating a millionth of a millionth.

Pigtail

A short length of fiber with a connector at one end and bare fiber at the other.

Primary coating
Buffer, buffer coating, primary buffer. A plastic coating applied to the cladding during manufacture to provide mechanical protection.

Ray
A line drawn to represent the direction taken by the light energy at a point.

Rayleigh scatter
The scattering of light due to small inhomogeneous regions within the core.

Refraction
The bending of a light path due to a change in refractive index.

Refractive index
The ratio of the speed of light in a material compared to its speed in free space.

Regenerator
Placed at intervals along a digital transmission route, it reconstructs the digital pulses.

Repeater
A transmitter and a receiver used at intervals along a transmission route to increase the power in an attenuated signal.

Sheath
A plastic coating which covers one or more optic fibers. The first layer is called the primary sheath and the outer one, the secondary sheath. Beware. The terminology is not consistent between manufacturers.

Signal to noise ratio
The ratio of the signal level to the background noise. Usually measured in decibels.

Singlemode fiber
An optic fiber which propagates a single mode.

Skew ray
A ray that never passes through the axis of the core during propagation.

Spectral width
The range of wavelengths emitted by a light source.

Splice
A permanent means of connecting two fibers. Alternatives are fusion splice and mechanical splice.

Star coupler
A device that allows a single fiber to be connected to several others.

Step index fiber
A fiber in which the refractive index changes abruptly between the core and the cladding.

Tee coupler
A 1 x 2 coupler used to tap off a proportion of the power from a system. It usually has a high splitting ratio and is used as part of a network.

Threshold current
The lowest current that can be used to operate a laser.

Total internal reflection (TIR)
Reflection occurring when the light approaches a change in refractive index at an angle greater than the critical angle

Wavelength division multiplexing

The simultaneous transmission of several signals of different wavelengths along a single fiber.

WDM

Abbreviation for Wavelength division multiplexing.

Windows

Commonly used bands of wavelengths. The 1st window is 850 nm, the 2nd is 1300 nm and the 3rd is 1550 nm.

Quiz time answers

Quiz time 1

1 c

2 a

3 d

4 d

5 c

Quiz time 2

1 b

2 d

3 a

4 a

5 d

Solution to quiz time 2 question 5

Using Snell's law:

$$n_1 \sin\phi_1 = n_2 \sin\phi_2$$

Insert the known values:

$$1.49 \sin 50° = 1.475 \sin\phi_2$$

assuming ϕ_2 is the angle shown as ?.

Transpose for $\sin\phi_2$ by dividing both sides of the equation by 1.475. This gives us:

$$\frac{1.49\sin50°}{1.475} = \sin\phi_2$$

Simplify the left hand side:

$0.7738 = \sin\phi_2$

The angle is therefore given by:

$\phi_2 = \arcsin0.7738$

So:

$\phi_2 = 50.7$

Quiz time 0

1 a

2 b

3 d

4 d

5 c

Solution to quiz time 3 question 2

$$f = \frac{V}{\lambda} = \frac{3 \times 10^8}{660 \times 10^{-9}} = 4.54 \times 10^{14} \text{ Hz}$$

Quiz time 4

1 d

2 b

3 c

4 c

5 a

Solution to quiz time 4 question 2

Using Snell's law

$$n_{air}\sin15° = n_{core}\sin(\text{angle of refraction})$$

So:

$$1 \times 0.2588 = 1.47\sin(\text{angle of refraction})$$

So:

$$\frac{0.2588}{1.47} = \sin(\text{angle of refraction})$$

Thus:

$$\text{angle of refraction} = \sin^{-1}0.1760 = 10.137° = 10.14°$$

Solution to quiz time 4 question 5

$$NA = \sqrt{n_{core}^2 - n_{cladding}^2}$$

So:

$$NA = \sqrt{1.47^2 - 1.44^2} = \sqrt{0.0873} = 0.2955$$

So:

$$\text{angle of acceptance} = \sin^{-1}0.2955 = 17.1875° = 17.19°$$

Quiz time 5

1 d

2 b

3 c

4 b

5 c

Solution to Quiz time 5 question 1

$$\text{Power level in dBm} = 10\log\frac{50 \times 10^{-6}}{1 \times 10^{-3}} = 10\log 0.05$$

$$= 10 \times -1.3 = -13 \text{ dBm}$$

Solution to Quiz time 5 question 2

As the input power is –5 dBm and the output power is –26 dBm, the difference in power levels is 21 dBm. The 2 km length of fiber has a total loss of:

$$2 \times 3 = 6 \text{ dB}$$

The known losses are the two attenuators and the fiber, so:

$$\text{total losses} = 2 + 4 + 6 = 12 \text{ dB}$$

As the total loss of the system is 21 dB, and we have only accounted for 12 dB, link A must represent the remaining 9 dB. At 3 dB loss per kilometer, this means that link A must be 3 km in length.

Solution to Quiz time 5 question 3

There are two methods.

The first is to convert the 0.25 mW into dBm then work in decibels and finally convert the answer back to watts.

The second is to convert the —15 dB into a ratio and work in power levels throughout.

The second method is slightly better as it only involves only one conversion:

$$-15 \text{ dB} = 10\log\frac{\text{power}_{\text{out}}}{\text{power}_{\text{in}}} = 10\log\frac{\text{power}_{\text{out}}}{0.25 \times 10^{-3}}$$

Divide by 10:

$$-1.5 \text{ dB} = \log\frac{\text{power}_{\text{out}}}{0.25 \times 10^{-3}}$$

Take the antilog:

$$0.0316 = \frac{\text{power}_{\text{out}}}{0.25 \times 10^{-3}}$$

So:

$$\text{power}_{\text{out}} = 0.0316 \times 0.25 \times 10^{-3} = 7.9 \text{ }\mu\text{W}$$

Solution to Quiz time 5 question 4

Loss is measured in dB, so that has eliminated two out of the four options.

When the figures are fed into the standard formula, the result is a negative value. This is because the output is less than the input:

$$\text{power gain} = 10\log\frac{\text{power}_{\text{out}}}{\text{power}_{\text{in}}} = 10\log\frac{0.8 \times 10^{-3}}{2 \times 10^{-3}}$$

So:

$$\text{power gain} = 10\log(0.4) = 10 \times -0.398 = -3.98 \text{ dB}$$

So there is a loss of 3.98 dB.

Quiz time 6

1 a

2 b

3 d

4 a

5 b

Solution to Quiz time 6 question 2

Starting with the standard formula:

$$\text{reflected power} = \left(\frac{n_1 - n_2}{n_1 + n_2}\right)^2$$

where n_1 and n_2 are the refractive indices of the two materials. So:

$$\text{reflected power} = \left(\frac{1.45 - 1.0}{1.45 + 1.0}\right)^2 = \left(\frac{0.45}{2.45}\right)^2$$

Dividing out:

$$\text{reflected power} = (0.1837)^2 = 0.03375 = 3.375\%$$

Taking the light input as 100%, the light output is:

$$100 - 3.375 = 96.625\%$$

This loss can be expressed in dB by treating it as a normal input/output situation:

$$\text{decibels} = 10\log\frac{\text{power}_{\text{out}}}{\text{power}_{\text{in}}}$$
$$= 10\log\frac{96.625}{100}$$
$$= 10\log 0.96625$$
$$= 10 \times -0.0149$$
$$= -0.149 \text{ dB}$$

So:

$$\text{loss} = 0.149 \text{ dB}$$

The minus disappears because it is referred to as loss.

Quiz time 7

1 c

2 d

3 b

4 a

5 a

Solution to Quiz time 7 question 2

Find the numerical aperture:

$$NA = \sqrt{(1.48)^2 - (1.46)^2}$$

So:

$$NA = 0.2424$$

Insert the figures into the formula:

$$\text{number of modes} = \frac{\left(\text{diameter of core} \times NA \times \dfrac{\pi}{\lambda}\right)^2}{2}$$

where NA = fiber numerical aperture and λ = wavelength of the light source.

So:

$$\text{number of modes} = \frac{\left(62.5 \times 10^{-6} \times 0.2424 \times \dfrac{\pi}{865 \times 10^{-9}}\right)^2}{2}$$

$$= 1513.78$$

The number of modes cannot be a fraction so the calculated value is always rounded down to a whole number. Hence the answer is 1513.

Quiz time 8

1 d

2 a

3 c

4 d

5 a

Quiz time 9

1 b

2 b

3 d

4 c

5 a

Quiz time 10

1 c

2 a

3 d

4 c

5 d

Quiz time 11

1 a

2 b

3 c

4 b

5 c

Quiz time 12

1 c

2 b

3 a

4 d

5 a

Quiz time 13

1 d

2 a

3 b

4 c

5 d

Solution to Quiz time 13 question 2

Start by converting the excess loss of 0.4 dB into a power ratio to find the total output power available:

$$-0.4 = 10\log\left(\frac{power_{out}}{power_{in}}\right)$$

Divide both sides by 10:

$$-0.04 = \log\left(\frac{power_{out}}{power_{in}}\right)$$

Take the antilog of each side:

$$0.912 = \frac{power_{out}}{power_{in}}$$

Insert the known value for the input power:

$$0.912 = \frac{power_{out}}{0.2 \text{ mW}}$$

Transpose to find the output power:

$$power_{out} = 0.912 \times 0.2 \text{ mW} = 182 \text{ } \mu W$$

This is the power level within the coupler after the excess loss has been taken into account. It is now split in the ratio of 9:1 so the tap power is 0.1 of the available power:

$$\text{tap power} = 0.1 \times 182 = 18.2 \text{ } \mu W$$

Quiz time 14

1 b

2 b

3 b

4 d

5 d

Quiz time 15

1 a

2 d

3 b

4 c

5 a

Quiz time 16

1 c

2 a

3 c

4 c

5 a

Solution to Quiz time 16 question 3

$$\sqrt{0.5^2 + 1.5^2 + 0.025^2} = 1.58 \text{ ns}$$

Remember that the fiber rise time was in ps, and as we are working in ns, it had to be converted to 0.025 ns.

Solution to Quiz time 16 question 4

Step 1

The known losses.

We have two pairs of connectors at 0.2 dB each = 0.4 dB and a fusion splice of 0.1 dB.

The present losses amount to 0.1 + 0.4 = 0.5 dB.

Step 2

In addition, we must include the aging losses, repair losses and the 'spare' power held back for emergencies.

For repairs, the assumed value is 0.1 dB and for aging losses 1 dB in total.

Spare, 3 dB as usual.

So:

total losses = losses from Step 1 + repair power + aging loss + spare power

In this case:

total losses = 0.5 + 0.1 + 1.0 + 3.0 = 4.6 dB

Step 3

Find the power available for the required cable run.

Transmitter output power is quoted as 30 µW which needs to be converted into decibels:

$$\text{decibels} = 10\log\frac{30 \times 10^{-6}}{1 \times 10^{-3}} = -15.22 \text{ dBm}$$

The total power available for the system is the difference between the transmitter output power and the minimum input power of the receiver.

In this case it is 14.78 dB.

Of this 14.78 dB, we have already accounted for 4.6 dB.

The spare power for the fiber is 14.78 − 4.6 = 10.18 dB.

Step 4

The fiber in use has a loss of 3.2 dBkm^{-1}, so the length that could be accommodated is:

$$\frac{10.18}{3.2} = 3.18 \text{ km}$$

giving an approximate value of 3.2 km.

Solution to Quiz time 16 question 5

$$\frac{0.44}{44 \times 10^{-12}} = 10 \text{ GHz}$$

Quiz time 17

1 d

2 a

3 c

4 b

5 b

Index